MOVE!

Also by Caroline Williams

Override

MOVE!

The New Science of Body Over Mind

CAROLINE WILLIAMS

PROFILE BOOKS

First published in Great Britain in 2021 by
Profile Books Ltd
29 Cloth Fair
London
ECIA 7JQ
www.profilebooks.com

1 3 5 7 9 10 8 6 4 2

Typeset in Sabon by MacGuru Ltd
Printed and bound in Great Britain by Clays Ltd, Elcograf S.p.A.

The moral right of the author has been asserted.

A CIP catalogue record for this book is available from the British Library.

ISBN 978 1 78816 461 0
Export ISBN 978 1 78816 728 4
eISBN 978 1 78283 662 9
Audio ISBN 978 1 78283 846 3

For Jon and Sam,
who keep me moving

Contents

Introduction

'Just come in and move however your body wants to ...'

This is the moment I have been dreading all day. It's 7.30 p.m. on a Wednesday and I have come to a village hall in Surrey, England, for what I've been told will be an evening of mind-altering free-form dance.

The young man on the door takes my money and encourages me inside. It's dark, apart from a few candles and fairy lights, but I can just make out a middle-aged DJ with a bleached crew cut and harem pants, spinning what probably comes under the banner of gentle tribal beats. One woman is rolling on the floor, while another is wafting around chasing imaginary butterflies. Then two of them start hugging. At which point my body tells me very clearly that it would like to move, as quickly as possible, back through the door.

I don't, and as the evening progresses, my body gives up the fight and starts to move. As the drums build towards a climax, the DJ murmurs 'let go' into the mic. Suddenly, as if he flicked a switch, I notice that I'm no longer moving my legs: they are moving me. My feet are hammering the floor at an alarming rate, as my head shakes from side to side and my arms flail in circles. I couldn't stop if I wanted to, and I feel unleashed, alive, free.

It was my first foray into the world of how movement

can radically affect the mind, and it was something of an eye-opener. I'll be honest: getting high on life like this isn't my usual kind of thing at all. My thing is more about sitting quietly, reading, thinking and writing about the peculiarities of the human mind – trying to get my head around why people think the way we do, and what science can tell us about how we might overcome the many mental and emotional glitches – from a lack of focus to anxiety and depression – that seem to come as part of the deal.

But then one day it occurred to me that my mind seems to work best when my body is on the move, and I started to wonder why. What is it about going for a long walk that unravels tricky scientific concepts in my head and helps a jumble of ideas finally coalesce into sentences? Why does an hour of yoga make me feel calm and in control, no matter what challenges the rest of the day holds? And why does jumping around to music in the kitchen make me feel so damn happy?

A spell of sitting and reading later, it transpired that I wasn't the only one asking these questions. Scientists working across a huge variety of disciplines, from neuroscience to cell biology, from exercise physiology to evolutionary biology, have started to investigate how bodily movement affects the mind and are just beginning to tease apart the physiological mechanisms that explain why. What they are finding is potentially game-changing scientifically and, given the way most of us live our lives these days, profoundly important for our overall health and well-being.

It shouldn't be news by now that most of us aren't moving anywhere near enough, myself included. After walking the dog for an hour in the morning, my working day mostly involves sitting at a desk and moving no further than the kitchen for multiple cups of tea. If he's lucky, the dog gets another wander through the woods, and on some days I do yoga, but more often than not weekday evenings involve yet more sitting, followed by eight hours in bed. Statistically, my life isn't all that unusual. The average modern adult spends 70 per cent of their life sitting or lying completely still; we move around 30 per cent less than our counterparts in the 1960s. Children spend up to 50 per cent of their free time sitting around, and that's before you factor all those hours bent over a school desk.[1] Elderly people, unsurprisingly, clock up even more hours of stillness, spending up to 80 per cent of their waking day barely moving a muscle.[2]

There are good reasons why we, as a species, have chosen the way of the sloth. First, it's comfortable. And second, we have spent most of the past century inventing technologies that make movement unnecessary. Unlike almost all of the other creatures on the planet, we are now in a position where we barely need to shift our bodies to find food, stay entertained or even find mates. Most of it can be done sitting down and occasionally moving our thumbs.

Yet while we (stiffly) pat ourselves on the back for having the brains to make this happen, we are missing something important. The brain evolved not for us to think but to allow us to *move* – away from danger and towards rewards. Everything else, from our senses to our memories, emotions and ability to plan ahead, was bolted on later to make these movements better informed. Moving is at the heart of the way we think and feel. If we stay still, our cognitive and emotional abilities become seriously compromised.

Sure enough, as we make ourselves comfortable, the

cracks in our collective psyche are beginning to show. Our increasingly sedentary lifestyles have been linked to falling IQs,[3] a vacuum in new creative ideas,[4] a rise in antisocial behaviour[5] and an epidemic of mental illness that is affecting people of all ages and from all walks of life.[6]

Studies suggest that both self-esteem and pro-social behaviour tend to be lower among people who spend more time sitting,[7] and that sedentary time is linked to a greater risk of anxiety and depression. Although it isn't clear which comes first, the sitting or the depression, physical activity is well known to be helpful in relieving symptoms of both conditions, so it stands to reason that a sedentary lifestyle is not ideal for anyone at risk of, or already dealing with, mental health issues.

Cognitive skills, too, take a hit when we sit. Being sedentary for long periods is the enemy of focused attention, memory and planning ability and puts an unnecessary lid on our creativity. A recent study of young Finnish schoolchildren found a significant relationship between the amount of time spent sitting and achievement in standardised maths and English tests over the course of two years, particularly among boys.[8] The rot sets in at an early age, and if we do nothing about it, sitting still becomes a lifelong habit.[9]

It also makes us old before our time. In studies, middle-aged people who spent more than two or three hours per day sitting in the car or in front of the TV were found to lose their mental sharpness significantly faster than those who were more active. We also know that regular exercise can reduce the lifetime risk of dementia by 28 per cent.[10] By one recent estimate, as many as 13 per cent of Alzheimer's cases worldwide can be traced back to a sedentary lifestyle.

Introduction

By another, reducing sitting time by a quarter could prevent more than a million new diagnoses worldwide. It doesn't matter how you cut the numbers, the message is the same: move more, and your brain will thank you in the long run.

Given our collective propensity for lolling around, it is perhaps more alarming than surprising that our sedentary ways may be affecting IQ at a population level, making humankind as a whole just a little bit less smart. IQ scores had, until recently, been rising by an average of 3 points per decade, for as long as people had been taking IQ tests and in countries all over the world. The trend is named the Flynn Effect, after the New Zealand-based psychologist James Flynn, who first documented it back in the 1980s.[11]

Soon afterwards, however, from the mid-1990s onwards, the Flynn Effect started to slow down, and by the early 2000s the trend was running in reverse, at the rate of a few points per decade.[12] Some observers explained this with the controversial claim that less intelligent people tend to have more children, which over time could have diluted national averages.[13] Others suggested that an increase in global migration was to blame, because incoming foreigners didn't quite understand the questions.[14] A recent study from Norway, though, shows quite clearly that neither of these – let's be honest, offensive – explanations holds water. By tracking IQ scores of young men from the same family over several decades, researchers found that IQ scores were declining within families across different generations. This means that it can't be down to changes in genetic fitness – evolution doesn't work that fast, and certainly not for complex traits such as intelligence, where variation is explained not by one gene but many. It is much more likely that the changes can be explained by changes in the environment.

Or, perhaps, by the way we choose to use it.

A lack of movement isn't the only change in our life-styles in recent years, but there is no doubt that the descent to seatedness is an important social change that has been creeping up on us for some time, and not only in the pampered West. A study from 2012 compared the amount of physical activity involved in work, leisure, home life and travel in the US, UK, China, India and Brazil from the 1960s onwards. Everywhere they looked, physical activity was on a downturn – and not only in leisure time but all the time. The fastest declines were in China and Brazil during the 1990s. These were mostly accounted for by changes at work and home, as physical labour gave way to office work and home appliances made daily chores less of a workout. Only India seemed to be bucking the trend, at least in 2012, but sedentary time had already shown signs of rising there too.[15]

The gym is not enough

If you're the kind of person who diligently hits the gym every day, you are probably feeling pretty smug right now. But there's a catch: exercise – at least the way we currently think of it, as something to do in earnest between long periods of sitting – is not the way to turn things around. Brain imaging studies show that there is a correlation between the thickness of brain regions involved in memory and the amount of time a person spends sitting, regardless of whether or not they also do high-intensity exercise at some point during the day. And while mood and focus do spike briefly after a period of exercise, overall it doesn't matter if you go for broke for an hour in your lunchtime spinning class. The effects of sitting still for the four hours either side of lunch don't go away.

In fact, you could argue that binge-exercising is missing the point of movement entirely. In her book *Move Your DNA*, movement guru Katy Bowman makes exactly this point. She says that exercising in short bursts, or to target certain muscles, is a bit like taking vitamin supplements to try and offset an unhealthy diet. It may help, but it's never going to make you truly healthy, and it will probably leave you hungry for what she calls 'nutritious movement'. Bowman doesn't delve all that deeply into how movement affects the mind, but I would argue that movement nutrition is at least as important for our mental, cognitive and emotional health as it is for our physical well-being. Moving our bodies in certain characteristically human ways connects us to the equally fundamentally human ways that we think, feel and make sense of the world that's both around us and within.

This is a theme that I'll return to in a few pages' time. For now, though, the important thing to know is that, as a society, we're not moving enough, and what little we are doing we are mostly doing wrong. That's the bad news. And here comes the good: it doesn't matter what you want to do with your mind. Whether you want to learn better, slow brain ageing, spark new ideas or just feel more in control of your mental health, there is almost nothing that moving more – and in particular ways – can't help you to achieve. Body movements can serve as a short-cut to changing the ways we think and feel.

This is a big deal: contrary to received wisdom, thoughts

don't only come from inside our heads, and thoughts are not the only routes to emotions. Some kinds of body movement help to reduce inflammation – a modern-day scourge linked to everything from depression to chronic pain. Others hijack the brain–body stress pathways in ways that help dial down feelings of anxiety and instil a visceral sense of confidence. Others change the way that electrical information flows through the brain, directly affecting our mental state. Move right and the body becomes an extension to and an equal partner of the brain – not just the meat suit that carries it around.

I say this with confidence because many scientists are changing the way they think of the body and its relationship to the mind. After many years in which science cast the body as an understudy in the story of our mental lives, it has finally landed itself a starring role. For decades the mind has been thought of as being run exclusively by the brain, sitting high and dry in the head while the squelching, churning, pumping and filtering activities of the body managed the dirty business of keeping us alive. Now, though, we now know that, unglamorous as they may seem compared with the electrical fizzing and whizzing of the brain, our bodily functions are just as much of a key component in what makes us tick.

As we'll see in the next section, the dirty business of keeping us alive actually involves a huge amount of communication between the body's various organs, the pipes and wires joining them up and the bodily fluids that swoosh around and between them. This communication provides a constant backing track to our lives, directing our thoughts and colouring the way we feel.

Introduction

In this new view the brain has a different, although no less important, role. According to the influential psychologist Guy Claxton, rather than being the master controller and arbiter of our every thought and decision, it instead acts as a kind of 'chatroom' that hosts the body–mind conversation that makes up our mental life. Here, he says, 'swarms of factors can come together and, through communication, agree on a plan'.[16] The brain is not so much the boss as a facilitator, bringing the key players to the table and allowing everyone to be heard and to come up with a collective plan of action.

'Action' is the important word here because that's where the link to movement comes in. The power of movement is that it allows us to hack into this body–mind chatroom and changes the tone of the conversation for the better. The overall aim of this book is to reveal some of these emerging dials and levers and how they work, using the best, most up-to-date science that we have.

In the pages that follow I'll meet not only the scientists who are investigating the physiological, neural and hormonal connections that link body and mind but also the many inspirational people who are putting the theory into practice and proving its worth in real life. From a psychologist who overcame illiteracy with the help of dance to an ultra-marathon runner who outpaced his demons, and from a neuroscientist who has found out he was wrong about Pilates to a stuntman who is helping kids backflip their way to better mind control, there is no shortage of people who are living proof of the movement solution. Science provides the data, but these people provide the inspiration to make a few simple changes that really can improve your life.

Move!

Ultimately, whether you are looking for more brain power, to feel more connected to others or just more in control of your life, the message from all corners of science is coming through loud and clear: this is no time to be sitting around.

1

Why We Move

*That which we call thinking is the
evolutionary internalisation of movement.*
Rodolfo Llinás

*Nothing in biology makes sense except
in the light of evolution.*
Theodosius Dobzhansky

Some days, the life of a sea squirt sounds almost idyllic. After a brief swim around the ocean while it's young and still has the energy, the tadpole-like larva finds a rock with a view and settles down for a rest. Once attached, it sets about developing into its adult form, a blob with two tubes. There it will sit for the rest of its life, gently sucking in water through one tube and blowing it out of the other like a small, rubbery bagpipe.

There's a high price to pay for this lifetime of relaxation. In its larval stage a sea squirt has a simple brain and a basic nerve cord that runs along the length of its tail. It uses these to swim around, searching for a good spot to live, and to co-ordinate its movements to get there. Once attached, though, glued firmly to the rock by its head, the sea squirt digests almost its entire nervous system, never to make a decision again.

The curious case of the disposable brain tells us something about why we have a nervous system at all. And, before we get to the 'how to' aspects of body movement over mind, it's worth considering why the many body–brain pathways came to exist. The distinguished Colombian neuroscientist Rodolfo Llinás used the sea squirt to make the case that animals originally evolved brains not so that they could *think* but so that they could *move* – away from danger and towards where the living is easier, making informed decisions as they go. Movement, Llinás reasoned, is simply too dangerous to attempt without a plan.[1]

Sea squirts represent a snapshot of a time in evolution when life was experimenting with whether a nervous system was any more likely to make you survive the rigours of existence. Nervous systems are expensive to run – our own brains gobble up 20 per cent of our body's entire energy budget despite accounting for only 2 per cent of our body weight. For the sea squirt, the answer was that the investment was worth it for as long as it was on the move, but thereafter, not so much. And when movement is no longer necessary, thinking is surplus to requirements, so the whole system goes in the recycling.

Since this period of evolutionary dithering, most species of animals have opted not only to keep a brain throughout the entirety of their lifespan but also to invest heavily in its architecture. Thinking and movement have been evolving in lockstep ever since. The human brain is by no means the pinnacle of the process of brain development – each creature's brain is, after all, uniquely adapted to its own way of life – but in terms of investment it is certainly an extreme example. Our brains contain three times as many neurons as our closest living relatives the chimpanzees. In fact, with 86 billion neurons with over 100 trillion connections between them, the human brain is the most complicated object we have ever encountered.

Explanations of how we got this way generally concentrate on our cortex, the wrinkly outer layer of the brain, which is disproportionately large in humans compared with other apes. The wrinkles are actually a product of its size: as the cortex expanded, adding more and more processing power, the only way it could fit into the skull was to repeatedly fold in on itself. Other species with a smaller cortex, such as dogs, cats and chimpanzees, have far fewer folds and wrinkles than us. Some, including mice, rats and marmosets, don't have any at all – their brains are as smooth as raw, skinned chicken.

Some think that our cortex enlarged to cope with the challenges of finding new ways to think – keeping track of our complex social lives, for example, or predicting where the next meal might show up and working out how to catch it. Then, once we used our big brains to work out how to cook, they got even bigger, because cooking allowed us to extract more calories from our food. All of this added up to an unusually large cortex that allows us to plan, to mentally travel back and forward in time and to come up with ideas for things that have never existed before.

That's a good synopsis as far as it goes, yet it totally ignores the influence of movement. A new theory adds this vital detail into our origin story, linking the evolution of forward thinking not to abstract computations inside the head but to a growing evolutionary pressure to come up with new ways to move. In this view, the origin of our most impressive mental tricks can be traced even further back in our evolutionary history, to a time before humans existed, when our even more distant ancestors needed to find new ways to get around.

Twenty-five million years ago the common ancestor that we share with other apes split off the evolutionary tree from the monkeys. These early apes lived in the trees like their monkey cousins but were bigger, heavier and clumsier and were in constant danger of falling from the branches. Their solution to this problem was quite sensible: to spend more time bearing their weight on their hands, holding on tightly to branches above in situations where smaller monkeys might have been able to balance. This strategy worked well, and over millions of years (and some shoulder modifications) slowly evolved into an ability to brachiate – swinging arm over arm in the trees at speed, as gibbons do today.

Brachiation is a complicated way of moving. According to the evolutionary anthropologist Robert Barton, of Durham University, it requires more than just a vague plan of action to stand any chance of getting from A to B safely. Staying safe while swinging through the trees requires an ability to link movement to an understanding of the consequences of your actions at speed – *I put my hand here, swing and reach … that one won't hold my weight, so I'll grab here* and so on – which means being able to formulate and adapt a plan on the fly. In a paper published in 2014 Barton put forward his idea that the development of the extra brain circuitry needed to support this new skill not only led to an improvement in our ancestors' physical gymnastic skills but also set the stage for our impressive mental gymnastics.[2]

The circuitry that is in charge of these kinds of super-fast movements is found not in the wrinkly cortex but in the cerebellum – the small, cauliflower-like region that, in diagrams at least, looks as if it's dangling from the bottom of the rest of the brain. At about the time that the early apes started swinging through the trees, the cerebellum started expanding, becoming disproportionately large compared with the cortex. This trend continued through the evolution of the great apes and accelerated in the branch that led to us.

The way the cerebellum is built seems to have made this expansion a fairly straightforward process. While the rest of the brain's wiring looks a bit like the organised chaos of an old-fashioned telephone exchange, the cerebellum is more like a well-kept vineyard, with neat rows of neurons linked with super-fast input and output wires. That means that another 'module' can be replicated and then bolted on fairly quickly, at least on an evolutionary timescale.

Until recently this finding would have raised a huge 'so what?' in evolutionary biology circles. The cerebellum had long been known to be specialised for fine-movement control. It shouldn't have been terribly surprising that the cerebellum would expand to support a complex new movement skill.

Then in the late 1990s and early 2000s the view of the cerebellum started to change. It was gradually becoming clear that what the cerebellum does for movement it also does for thinking and emotional control. Brain imaging experiments and tracing of neurons throughout the brain revealed that many of the evolutionarily newer cerebellum 'modules' wire up to the frontal parts of the cortex, which are in charge of planning and forward thinking and help to fine-tune our emotional reactions. In fact, it turned out that only a small portion of the human cerebellum connects to the movement-generating parts of the rest of the brain. The rest specialises in thinking and feeling.

Barton's theory is that, when brachiation tied together movement, forward planning and potentially fear of falling from a great height, it set us up for all manner of sequential thinking, from understanding the rules of language and numbers to building simple tools, telling stories and working out how to get to the moon and back. It's tempting to speculate that it may also underlie the sensations that accompany some of our less successful social interactions: swinging and falling is certainly how it can feel when a conversation suddenly takes a turn for the worse.

The ability to think sequentially is particularly useful for skills that require not only fine sensory motor control but also a capacity to work out a sequence of actions that will lead you to your goal – central to the ability to knit a scarf or think through a series of moves in chess. It could also explain how chimpanzees can work out the sequence of movements that will allow them to adapt a twig to fish for termites. 'Our capacity to work out how to achieve a goal by stringing together a sequence of actions is kind of the basis of our causal understanding of the world,' says Barton.

Blame the ancestors

Twig technology aside, the other great apes haven't done a great deal with their expanded forward-planning skills.

Humans, though, took the ball and ran with it in a big way. One potential reason for this is that when our ancestors split off from the other apes, they started adopting a very different lifestyle, one in which they spent far less time in the trees and started roaming longer distances on the ground in search of food. The mental and physical demands of this new lifestyle brought about another crunch point in evolution, where new ways of moving and thinking came together and worked hand in hand to increase the species' chances of survival. As a result, being physically active started to become non-negotiable to keep the brain working at peak capacity.

It's worth a quick aside at this point to remember that evolution doesn't work with an endgame in mind. Our minds and bodies didn't get to be the way they are today because evolution had a plan to make us the cleverest, most self-aware species on the planet. We got here because the changes that brought us to this point must have provided some kind of survival advantage when they first appeared. Each of them had to be useful from the start, and they stuck around because they continued to supply benefits.

Use it or lose it, then, is a rule of evolution in general, but in terms of our physiological responses to movement it applies to us especially. It's well known that our ability to exercise – our cardiovascular fitness, our muscle strength and so on – is directly linked to how much we as individuals have challenged those systems in the past. That isn't the case for all species: bar-headed geese, for example, manage a 3,000-kilometre migration each year with no training at all. The physiological changes that build them stronger flight muscles and a bigger, more efficient heart are triggered not by months of intense training but by the change of season and a lot of extra food.[3] It's the stuff of dreams – imagine if the shortest day heralded not only the coming spring but also an increasingly fit and toned beach body, just in time for summer (but only if you ate enough pizza).

Unfortunately, our bodies aren't built that way, and it seems that the same 'use it or lose it' rules apply to the brain. According to David Raichlen, who studies human evolution at the University of Southern California, this is a feature that can be traced to a point in time, around 4 million years ago, when our ancestors stopped being ape-like animals who sat around in the trees all day eating fruit, and started to explore.

At the time the climate in East Africa was becoming cooler and drier, and tropical forest was giving way to woodland and savannah. This made food more difficult to find and forced our ancestors to forage further afield. Under these circumstances, evolution would have favoured those who could stand up straighter to walk or run long distances in search of food.[4]

Those who were not only able to walk and run long distances but also to make intelligent decisions – finding their way to where the best forage is found, remembering the way back to base and so on – were even more likely to survive and pass on their genes. Around 2.6 million years ago, when hunting skills were added to gathering, thinking on our feet became yet more critical. Now our ancestors not only had to forage widely and wisely but also had to work together to outwit and bring down larger prey. And so these two selection pressures – to walk further and think better – were tied together in the unique evolutionary history of our species.

As a result, says Raichlen, our physiology became fixed so that, when we exercise, the brain responds by physically adding more capacity.[5] The hippocampus, a part of the brain that is involved in spatial navigation and memory, responds to physical exercise by adding new cells – essentially adding capacity to the brain's memory banks. If this new capacity is then called upon in future foraging or hunting bouts, it is more likely to be retained. New neurons are only part of this brain-boosting process. The extra capacity also requires more blood vessels, which allow for more fuel and oxygen to flow around the brain, helping it to do its job.

On the flip side, if the new memory banks are left idle, the brain will begin to make energy savings, removing any architecture that isn't strictly necessary and trimming unused capacity to claw back some of its energy budget and divert it to where it's needed.

The upshot of all this is that, while our closest relatives among the great apes get away with being couch potatoes, moving only if they can't possibly avoid it and suffering no physical or mental repercussions for their laziness, we, like the sea squirt before us, can't. The specific challenges of survival as a hunter–gatherer tied the nuts and bolts of our mental capacity to our levels of activity.

Sitting around is no longer an option for humankind if we want a healthy body and mind: that ship sailed when our ancestors gave up a life of fruit in the trees. As for how much we need to move, studies of the Hadza people, modern hunter–gatherers who live in northern Tanzania, have found that women walk almost 6 kilometres per day, while men cover 11.5 kilometres, the equivalent of 8,000–15,000 steps. If we take this as a rough guide to what our bodies evolved to do, it means that getting your steps in is non-negotiable for a fully functioning brain. If you don't like it, take it up with *Homo erectus,* the species of ancient humans that started the whole sorry business.

On the upside, the evolutionary pressures that link moving and thinking are the very same ones that make moving feel good – including the well-known endorphin boost, which makes exercise feel effortless, even euphoric, and encourages us to keep going when we start to get tired. On the other hand, it raises the worrying possibility that, if our minds are there to help us move – and we don't – perhaps we risk a future as a race of couch-bound filter-feeders, our hard-earned brains turned to mush.

It's too soon to panic, though. Humans are nothing if not adaptable. What we need to do is use that adaptability to spring into action once again, to unglue ourselves from the sofa, get up and remember how good it feels to move.

Travelling without moving

The final part of our moving, thinking and feeling story is more difficult to pin down to a particular point in our evolutionary history, not least because we can't see it happening in our own heads, let alone those of other species. But we do know that it must have happened, because at some point we became able to move not just physically but also virtually, inside our mind's eye.

Whether other species can do this too is very much a moot point. There is some evidence of what looks a lot like thinking ahead in some species. In 2009 a captive chimpanzee called Santino was seen calmly piling up rocks in his compound at

Furuvik Zoo in Sweden, which he would later hurl at visitors in what looked a lot like a premeditated attack.[6] Similarly scrub jays, one of the cleverest members of the crow family, cache food to eat later. In experiments where they were fed boring kibble and then occasionally given something more exciting, they seemed to plan ahead and store more of the good stuff for later, when plain rations would presumably return.[7] While some call this evidence of forward thinking, other scientists insist that it doesn't prove that they are preparing for the needs of their future selves. Until we find a way to talk to the animals, we will never know for sure.

We do know, however, that humans definitely can relive the past and plan for the future. The ability to imagine things that have never been, mentally to travel back and forth in time to learn from the past and to plan for the future is very much a human speciality, and it all comes down to what Rodolfo Llinás calls the 'evolutionary internalisation of movement'. From Llinás's point of view, thinking and moving are basically the same kind of thing. The only difference is that movement has a final stage that makes it real to the outside world too.

The advantages of this ability are obvious. Unlike moving, thinking is invisible and risk-free, allowing us to explore the world in our own minds, trying things out for size and updating them based on new information before we risk life and limb. Something similar is true for emotions. The whole point of emotions is to stir us into action to change something that isn't right in the world: the word 'emotion' comes from the Latin for 'to move away'. It stands to reason that if the process of moving can begin mentally, before it shows on the outside, that gives an animal a huge advantage in terms of outwitting predators or rivals and navigating our complex social world.

Interestingly, experiments dating back to the 1960s showed that the body–brain–understanding system has to be trained on real-life movement if it is to work later in the virtual world of our minds. In a classic (but heart-breaking) experiment into visual perception two kittens were strapped into a kitten-sized carousel.[8] There they would spend their days going round and round, each looking out at exactly the same laboratory-based view as the other. The only difference between them was that one kitten had its feet on the floor and could drive the carousel by walking forwards. The other was suspended in a box, with no contact with the ground or control over the carousel's spin. After a few weeks like this the kittens were finally freed. The one that was allowed to drive the carousel with its feet was seemingly fine – able to see normally and move through the world with no problems at all. The other one was, to all intents and purposes, blind: it couldn't avoid obstacles and couldn't navigate the room safely. The scientists concluded that because the kitten hadn't been able to link its body movements with the changing outside world in early life, it never learned to make sense of what its eyes could see.

What it's like

Outside of the lab, of course, these connections between movement and internal experience happen automatically, and gradually build up to provide the basis for a rich understanding of our place in the world and how our actions affect what we experience.

This process may even explain a basic mystery of human consciousness: why we have such rich sensory experiences that exist only in our mind's eye. How is it that we can so vividly imagine smelling a rose or seeing a sunset, for example, or conjure up the warm and fuzzy feeling of hugging someone we love? These imagined experiences feel like they are in our heads, but philosopher J. Kevin O'Regan of Paris Descartes University points out that these experiences begin with the way we move our bodies and physically interact with the environment.[9] These sensations then get unhooked from bodily experiences and become amplified, going around and around in a mental loop, becoming more intense as they go. According to this theory, our rich imagination – the ability to 'feel' the sensations in a piece of writing or be 'moved' by a piece of art – comes from the way that our movements and interactions with the world can be detached from the world outside and sent undercover, where we can enjoy them privately.[10]

In sum, whether it's our capacity to plan ahead, to remember where we are and what we are doing, to imagine the future or to feel deeply, the very experience of being human is intimately tied to our movements through the world. Vital, in fact, to the very concept of the mind.

Mind in body or body in mind?

This seems like a good point to acknowledge that the ideas in this book are tied up with some fairly hefty ongoing scientific and philosophical debates, the biggest being what – and where – the mind actually is.

In the view of cognitive scientists the mind is a construct

of the brain. According to this argument, the brain functions as a kind of master computer, with the neurons and other cells of the nervous system acting as the hardware on which the software of the mind runs. From this point of view, the body is important, but mostly as a source of inputs into the system. It's up to the brain's clever algorithms to work out what's going on and decide what to do about it.

The concept of the body doing the bidding of the all-powerful brain is probably how most people think about things. It's even reflected in popular culture: in the classic 1990s film *The Matrix* intelligent machines grow humans in vats, keeping them busy with a fake version of reality directly projected into their brains. When Neo needs to learn Kung Fu – no problem, there's an app for that.

People who work in embodied cognition don't buy into this at all. They see the brain not as a master computer but as one node of a much larger network that spans not only the wider body but also the surrounding environment. In this view, it wouldn't matter how much Neo's brain knew about Kung Fu if he hadn't learned the movements by actually doing them. Like the poor kitten on the carousel, he'd have no hope of putting what he'd learned into action.

The body certainly knows more than we generally give it credit for. Thanks to our sense of 'proprioception', the implicit knowledge of where our body is in space, we can move around without banging into things, adjust our balance without thinking about it or reflexively stick a hand out to catch a ball that is about to hit us in the face. Through proprioception we instinctively know where we are, how we are moving and where our body begins and ends.

Then there is the more mysterious sense, interoception:

our ability to detect the internal physiological state of the body. All day and night the body busies itself tweaking countless physiological dials that keep our biology within a safe, liveable range. This constant tweaking, called homeostasis, is an ongoing effort with different systems managing their respective departments – heart rate, blood sugar level, water balance and so on – while keeping each other abreast of any news. Some of these changes we are conscious of (a racing heart, for example), and others we are not. Nevertheless, according to the Portuguese neuroscientist Antonio Damasio, of the University of Southern California, they influence our minds all the same.

For Damasio, the ongoing process of homeostasis, whether conscious or unconscious, is a central building block of our sense of self and of how that sense of 'me' experiences the here and now. Through homeostasis, and through our interoceptive sense of what is happening, we know whether we are on edge or relaxed, tired, thirsty or in need of a snack. Interoceptive ability varies, and the better a person is at sensing their internal state, the more likely they are to take action to put things back into balance – to seek out rest or to get away from someone who gives them a bad 'gut feeling', for example.

This isn't to say that the brain isn't involved – clearly it does play an important role in our mental lives. But in the embodied view the brain isn't there to give orders: it's there to pull together the strands of our internal experience so that the system as a whole can make sense of them. The insula, an area of cortex that is found deep in one of the folds of the brain, just above each ear, seems to play an especially important role in all of this, combining interoceptive messages with proprioceptive ones, and with information coming in through the senses, to come up with what the neuroscientist Bud Craig calls a 'global emotional moment' – a sense of 'how I feel right now'.[11]

Where are you?

Of course, none of this helps to settle the argument about *what* the conscious mind is actually made of, where it is and what it would look like if you wanted to point and stare at it. Back in the seventeenth century the French philosopher René Descartes famously threw his hands up and declared that while the body (including the brain) is a physical thing, the mind is made of something else entirely, something that is both invisible and immeasurable. This general consensus has prevailed ever since, not least because if the mind is made of 'stuff', we've yet to find a way to quantify it.

Many neuroscientists and philosophers – and the Buddhist scholars who have in fact been saying it for the longest – believe that what we think of as the mind is actually an illusion that comes as an accidental side-effect of the binding together of messages flying around the body and brain into one 'self'.

The embodied approach sees our conscious self as being grounded in, and bound together by, the sensory experiences of the body and its interactions with the world. In recent years neuroscientists have begun to put these things together and come up with a unified explanation: that the mind is the result of an ongoing process of predicting what is probably happening, both in the world outside and within our bodies, and then taking action to adjust the dials. Moving in the world and interacting with it are the best way to make sense of what the brain thinks is true.

And this is where the importance of movement comes in. Moving the body not only changes proprioception but can also have knock-on effects on information coming in from the senses, and on interoception, via changes in the internal state of the body. By changing the chemical and physical basis of how we feel, movement allows us to change the inputs to the 'global emotional moment', leaving us with a different sense of 'how *I* feel *now*'.

This, in a nutshell, is what the rest of this book is all about. As we'll see time and again, it is entirely possible to use the way we move as a form of self-management for better physical and mental functioning. And whether you believe that that you-ness of you lives in your head and looks out through your eyeballs, whether 'you' are distributed throughout your body, including the brain, or whether there is no you at all, it doesn't matter. The truth is that brain, body and mind are part of the same beautiful system. And the whole thing works better when it's on the move.

2

The Joy of Steps

All truly great thoughts are conceived by walking.
Friedrich Nietzsche, 1889

Charles Darwin had a lot of thinking to do. It was the summer of 1842, and already he'd been back from the voyages of the *Beagle* for more than five years. He had scribbled his first sketch of the tree of life almost as soon as he stepped ashore,[1] but what with the noise and bustle of London and a growing family at home, he could barely hear himself think, let alone formulate a revolutionary new theory of biology.

His solution was to move – and in more ways than one. First, he relocated his family to a quiet corner of the English countryside where the kids could play somewhere other than just outside his study. Once there, he set about constructing what he came to call his 'thinking path' – a quarter-of-a-mile gravel path around the grounds of his home, which passed by a rolling meadow before looping back through a dark patch of woodland. It was there, on his daily four or five circuits of the path, that Darwin finally found the headspace to come up with his theory of evolution.

Walking the thinking path today with my son and his friend trailing behind, giggling about something they saw on

YouTube, I feel Darwin's pain. But the emerging science of movement suggests that it was more than just simple peace and quiet that helped Darwin think more clearly. Walking is proving to be a multi-use mental tool, which can affect both our psychology and our physiology in very specific ways. These changes can, in turn, transform the way we think and feel.

That walking and thinking are connected is hardly head-line news. But while generations of geniuses, from Friedrich Nietzsche and Virginia Woolf to Bill Gates and Steve Jobs, have made the case for thinking on foot, we are only now discovering how and why it works so well. And, perhaps more importantly, science is beginning to reveal how different ways of doing it bring specific mental benefits, depending on what it is you are trying to achieve.

It might sound a little ridiculous: who actually needs to be told how to walk? But research coming from the fields of evolutionary biology, physiology and neuroscience is all pointing to the fact that walking a lot, and running a little, made our species what it is today. If we don't do it enough, we risk losing our mental and emotional edge. And with researchers linking everything from falling IQ to a lack of creative ideas and failing mental health to our sedentary life-style, there are plenty of reasons to relearn what we think we already know.

Fittingly enough, given Darwin's fondness for steps, the first evidence that walking and thinking are intimately con-nected comes from the story of the evolution of our species.

As we saw on p. 18, before the invention of hunting and gathering, our distant ancestors were basically layabouts who spent most of the day sitting around, munching fruit and perhaps the odd tuber. Like most of us today, they probably only averaged about 3,000–5,000 daily steps but, unlike us, they were none the worse for it because their physiology was fine-tuned to this level of fuel and activity.

Over time, though, the climate changed, woodlands turned to savannah and food became harder to find. Our ancestors had no choice but to roam further and wider to find enough to eat. Eventually some bright spark hit upon hunting and gathering as a way to gather enough calories to survive. In terms of survival this proved to be a good idea, which meant that evolution favoured those who were better adapted to walk and run long distances. We evolved to move and, like it or not, we all carry those genes today.

In 2017 David Raichlen, who studies human evolution at the University of Southern California, and his colleague Gene Alexander, at the University of Arizona, described this relationship with what they called the adaptive capacity model. It was the first time anyone had made the connection between our evolutionary history and the 'use it or lose it' plasticity of our adult brains. We've known for decades now that physical exercise is the best proven way to boost brain health and cognitive skills, including memory and attention, and to reduce the risk of depression and anxiety. Now there was a good reason why: we evolved to be, in Raichlen's words, 'cognitively engaged endurance athletes'.[2]

The 'cognitive' bit is important, because hunting and gathering is more than just physical work: you can't simply put one foot in front of the other and hope that something tasty crosses your path and lies down to be eaten – and with our comparatively weedy physical frame we can't rely on brute force to bring down big game. By necessity, human-style hunting is skilled mental work, which requires tracking and outwitting prey and predicting their next move, while working as a team, keeping an eye on the time, looking out for danger and remembering the way home. Gathering involves remembering where to find the good stuff and out-thinking other animals that want to eat you or steal your food.

As a result, our biological baseline is to be on our feet, moving and thinking at the same time. If we don't do it, our brains make the sensible decision to save energy by cutting brain capacity. In better news, when we get on our feet and move, it primes the brain to be alert and to learn.

A well-oiled machine

Putting this to use isn't actually all that hard. Evolution has built in several unique design features that link moving on our feet to a mental boost. And though most of us don't need to hunt and gather any more, the system still works just as well whatever it is you want to achieve.

Among the human-tailored design features are, of course, the usual suspects: the feel-good hormones, the endorphins and endocannabinoids, that are linked to the runner's high and the general feel-good factor that comes with exercise. Sure enough, studies have shown that we, along with other 'athletic' species, get a hit of the good stuff when we exercise. There are also Raichlen's experiments: he compared humans with dogs, which clearly love running, and ferrets, which aren't so keen. In terms of endocannabinoid signalling we have much more in common with dogs than with the more sedentary ferrets.[3] The downside is that walking isn't quite enough to bring on a high unless it leaves you seriously breathless. The feeling-great part only really kicks in after an intense run at a pace where it's difficult to hold a conversation.

Endorphins, though, are easier to come by – they tend to appear after just twenty minutes of brisk walking. Likewise brain-derived neurotrophic factor, or BDNF, a growth factor that not only enhances the growth of new neurons in the hippocampus, which is important for memory, particularly spatial memory, but also increases the likelihood that the brain will make new connections, boosting our ability to learn. Meanwhile, another growth factor, vascular endothelial growth factor (VEGF), gets busy adding new blood vessels to support the expansion.

These links are pretty well understood and now count as run-of-the-mill in exercise physiology and mental health. But there are a few new kids on the block that are perhaps more surprising.

Who, knew, for example, that our feet come supplied with a set of inbuilt 'pressure sensors' that work with our beating heart to send more blood to the brain? This was the finding of an engineer called Dick Greene, who, having spent many years working in the oil fields of Texas, decided in the 1970s to turn his attention to the pipework of the human body. Back then the received wisdom was that, even when the heart rate rises to send more blood to working muscles, the brain as a whole gets no more than usual because our blood vessels adjust their diameter to keep blood flow constant and shield the brain from dips and surges. It does this for a good reason: too little blood at any one time and the tissue could be starved of oxygen and die; too much and the brain could swell, squashing delicate neural tissue against the skull.

Greene, though, suspected that there might be more leeway in the brain's blood supply than conventional wisdom allowed. Using the technology of the time, however, it was only possible to measure blood flow to the brain while people were lying still, often with measurements being taken directly from the arteries and veins, so it was impossible to know whether moving changed anything. Then Greene worked out a way to measure blood flow in the carotid artery of the neck using non-invasive ultrasound mounted on a headset that allowed measurements to be taken continuously, even while his subjects were up on their feet and moving. As he suspected, he and others found that any form of aerobic exercise will increase blood flow to the brain by around 20–25 per cent, at least in the short term.

Crucially, though, he recently found that putting your full body weight on your feet while you are exercising provides an added boost. In 2017 Greene reported that putting weight on your feet compresses the major arteries of the feet, increasing turbulence in the blood and increasing blood flow to the brain by a further 10–15 per cent.

Whether this extra blood makes the brain work better in the moment or whether it's more of a long-term, oiling-of-the-cogs kind of effect is something Greene and his team are still working on. The studies he had planned for 2020, to measure blood pressure and flow in healthy people who are standing, walking and running, were shelved indefinitely because of the outbreak of COVID-19.

Intriguingly, though, he has found a sweet spot where the rhythm of our footsteps synchronises with our heart rate.

In Greene's experiments the biggest boost in blood flow happened when the heart rate and step rate synchronised at around 120 beats and steps per minute. Walking in sync with the heart rate seems to provide a steady and predictable increase in blood flow to the brain, which, Greene speculates, might contribute to the feel-good factor that comes from a good, brisk walk.

Perhaps unsurprisingly, an even bigger blood flow boost comes from running, when the feet are hammered with 4–5G of force with each step as you pound the ground. But, as Greene told me on a mid-hike video call from the mountains of Idaho, cushioned shoes probably take some of that benefit away. Running barefoot, or wearing minimal shoes, might well mean that you get a better boost, although this has yet to be confirmed in a scientific study.

All of this got me idly daydreaming about inventing a rhythmic foot massager that could boost blood flow to the brain and net me my first million. But other scientists make a good case for why actually standing up and moving are part of the deal.

In a word: gravity. Or, more specifically, the physiological changes that happen when we put weight on our bones, and what that, in turn, does to our minds.

We tend to think of bones as dry white sticks that hold up our insides, but in reality bone is a living tissue that is constantly being built up and broken down to adjust to the stresses that are – or are not – put upon it. We know this because, without the constant need to fight gravity to stay upright and move, astronauts and people who are bed-ridden for long periods quickly lose bone mass: the cells that break down excess bone work harder than those that build and repair. What's less well known is that the brain is also affected by a lack of bone-building. Studies have linked the loss of bone mass seen in osteoporosis to an increased risk of cognitive decline.[4] Astronauts also seem to suffer short-term cognitive problems after a stint in space, as do people who have experienced prolonged bed rest.

It is now emerging that these two things are very much connected by one strange and surprising fact: our bones are in constant conversation with our brains. What they talk about depends very much on how much we ask our bones to move while also resisting the pull of gravity.

To find out more I arranged to meet a bona fide legend of neuroscience, Eric Kandel, who won the Nobel Prize in Physiology or Medicine in 2000 for his part in the discovery of the molecular basis of how our brains store memories.

When we meet, on a bright October day in New York in 2019, he is a week away from his ninetieth birthday. He's still interested in memory but, perhaps unsurprisingly given his advanced age, has turned his focus towards maintaining memory well into later life. From what I can see, this seems to be going pretty well for him. He still works five days a week at Columbia University's shiny new Jerome L. Greene Science Center in West Harlem, and most days he walks the two and a half miles from home to his lab. His enthusiasm for science is as bright as ever, and he can't wait to tell me about his latest work on the link between movement and memory.

'I like to do a lot of walking,' he tells me. 'And reading about that, I came across the fact that bone is an endocrine gland and it releases a hormone called osteocalcin. I did some experiments where I put osteocalcin into experimental animals, and [found that] it enhances memory, and strengthens various intellectual functions. I said, gee, this is pretty nifty. I'm not wasting my time.'

The work he was reading about came from another Columbia University scientist, who is based a couple of miles north in the Department of Genetics and Development. Gerard Karsenty had been working on the genetics of bone since the 1990s, trying to work out why bones accumulate calcium and harden while other organs don't. The main candidate at the time was the gene for osteocalcin, a protein released only from osteoblasts – the cells that are responsible for building new bone. Since osteocalcin is released during the bone-building process, it seemed likely that it played a role in making bones physically strong.

In fact, as Karsenty told me when I went to his office to find out more, it was doing nothing of the sort. 'I was thinking that I would unravel the secret of bone mineralisation,' Karsenty recalls, looking both wistful and amused by his youthful ambition. 'And lo and behold, the bone couldn't care less whether there is osteocalcin or not.'

Even under an electron microscope the skeletons of mice that had been genetically engineered to lack osteocalcin looked entirely healthy. But it soon became clear that all was not well with the mice. For a start, they were unusually docile: they didn't try to run away from being handled or try to bite when they were picked up; they just sat there and let the world pass them by. Yet despite looking for all the world as if they were chilling out, they also showed more anxious behaviours than normal mice, being more likely to hide in a darkened corner than to explore somewhere new.

They also flunked a gold-standard test of mouse memory, the Morris water maze. In this test scientists first train mice to find a submerged platform in a deep and steep-sided pool. Once they have learned to find it reliably, the test is repeated but with murky water, to see if they can remember their way to safety. Healthy mice find this pretty easy, but the osteocalcin-deficient mice were clueless, swimming around aimlessly trial after trial. But when Karsenty injected osteocalcin into their blood, all of these problems went away, and they became as bright as the average mouse.

Twenty years of research in Karsenty's lab has since shown that osteocalcin is released during bone-building, not to strengthen us physically but to travel via the blood carrying messages to the brain. It does so via specialised receptors in the hippocampus, the brain region responsible for memory in general and spatial memory in particular. Without osteocalcin that communication doesn't happen, and in mice at least, the hippocampus and other brain regions end up smaller and less connected than normal.

Mice are not humans, obviously, but Karsenty is confident that these results apply to humans too. 'Bone is one of the latest organs to appear during evolution, and there are no genes that are expressed in bone in the mouse that have not been conserved in the human. So, it's unlikely that what we've seen in the mouse is misleading,' he says.

Only a few studies have so far been done in humans, but those that have been done suggest that there is a link between low osteocalcin levels in the blood and poor performance on cognitive tests from middle age onwards. One recent study found particularly low levels of osteocalcin in people with Alzheimer's disease. Both Kandel and Karsenty are independently doing further human studies, Karsenty into osteocalcin levels in neurodegenerative disease and

Kandel into the link between memory and variations in levels of osteocalcin circulating in the blood.

Depressingly for those of us of a certain age, the amount of osteocalcin in the blood peaks in early adulthood and starts to drop from around the age of thirty in women and forty-five in men. Kandel takes this as a sign that putting weight on our bones is essential at all ages, but particularly from middle age onwards. 'Movement is essential. And the older one gets, the more important it becomes,' he says.

One unknown factor is how much exercise is necessary to seriously up your osteocalcin levels. For his part, Karsenty isn't convinced that most of us are up to the challenge. 'Ideally, if you exercise every day since you are thirty years old, you would have more osteocalcin, but it's unlikely that anybody would do it,' he says. Plus, he says, osteocalcin only peaks for a couple of hours, before returning to an age-appropriate baseline. He suggests that an osteocalcin pill might be a better way to maintain memory, particularly in people who are less able to move.

But there's more than just memory at stake. Osteocalcin also talks to the muscles, telling them to release more fuel for exercise. In fact, it's starting to look like a multi-purpose hormone that tells the body it's time to think and move at the same time – more evidence that human beings are cognitively engaged athletes by nature. 'Movement is a survival function that requires muscle in running, but also knowing where to go, which is cognition. These functions are connected,' says Karsenty.

As for the question of why our bones evolved to specialise in memory and movement as well as scaffolding, Karsenty thinks that it's all part of an ingenious mind–body strategy that evolved to help us escape danger. In a recent series of experiments in mice, Karsenty's team showed that osteocalcin release from bone is a key part of the fight-or-flight response. When the brain signals danger, they found, osteocalcin is released from bone into the bloodstream, where it can circulate, turning off the 'rest and digest' part of the nervous system while revving up the body for escape.[5]

The memory boost that we get from osteocalcin is all about survival too. It helps us to remember the lessons of each emergency for next time. The bonus is that we can bypass the fear part of the process and choose to stress our bones instead: the mental benefits still appear.

As an aside, there may be yet another way to get your osteocalcin boost – no exercise or fear necessary. It has been known for some time now that blood from young mice has the power to boost both the health and brainpower of older mice. Off the back of this research, in 2016 a Silicon Valley start-up company called Ambrosia started selling blood transfusions from sixteen- to twenty-five-year-olds to people over the age of thirty, at $8,000 a pop.[6]

Between 2016 and 2018 the company ran an in-house clinical trial, which it claimed showed a decrease in blood markers for cancer, Alzheimer's disease and inflammation in thirty-somethings who received young blood. These claims have not been published in any scientific journal, and the methods used in the trial – participants paid $8,000 to participate, and there was no placebo group – attracted widespread criticism. In February 2019 the US Food and Drug Administration published a warning against plasma transfusions from private companies, stating that 'there have not been any well-controlled studies that show the clinical benefit of the administration of plasma from young donors, and there are associated safety risks'.[7] Ambrosia paused operations shortly afterwards before quietly returning in late 2019, offering transfusions from blood bank stock rather than direct from young donors.[8] With blood bank stock there's no guarantee that the blood comes from younger donors, however. According to the American Association of Blood Banks, the average blood donor in the US is between thirty and fifty years old and 16 per cent of donations are from over-sixty-fives.[9]

Back in the world of science, the head-scratching continues about what the secret ingredient in young blood might be, if indeed it does have the same effect in humans. Karsenty speculates that – in mice, at least – osteocalcin could be the answer. If you give old mice young blood without osteocalcin, the elixir of youth doesn't work its magic.

Does putting weight on your bones have the power to preserve memory and mood well into old age? And if some weight is good, would adding more, by adding ankle weights or carrying kettle bells, be even better? We don't know for certain. But, all things considered, it makes sense to not give into gravity but to fight it like your chance of a happy, healthy old age depends on it. Because there is a growing body of evidence that it might.

*

Leaving physiology aside for a moment, another reason why walking – and running – improves mental well-being is that it temporarily changes your window on the world. Whether walking, running or moving under your own steam by other means, there is no escaping the fact that you are, literally, getting somewhere. And this can tip over into a figurative sense of progress too.

Marcus Scotney stumbled on this principle about twenty-five years ago and, it's fair to say, took it to extremes. Having struggled with depression throughout his teens, he found that the only thing that made him feel better was running for the hills. And then over them. And then back down the other side. He turned out to be pretty good at it too – now, in his mid-forties, he is a professional ultra-marathon runner and coach who in 2017 won the Dragon's Back – a five-day, 188-mile race across the Welsh mountains that is known for breaking even the toughest of athletes. He did it in under forty hours, setting a new course record.

We arrange to meet in a car park deep in the UK's Peak District on a roasting hot day in August. It's a reunion of sorts – Marcus and I went to school together from the age of five until we were eighteen. We were never especially close friends, but there's solidarity among those who were too uncool for school (me too short and frizzy, him too skinny and ginger) that never quite leaves you, and we greet each other with a big hug.

I've made all the excuses I can think of to not do any running today but he's not having it. 'I need to interview you, Marcus,' I say. 'I can't run and talk at the same time.'

'If you're running so fast that you can't talk, then you're running too fast,' he replies.

'I'm not sure our stride lengths are compatible.'

'Actually, my stride length is quite short,' he says.

Luckily for me, on the day we meet he's tapering his training to prepare for the Ultra Trail du Mont Blanc, another of the toughest mountain marathons in the world, this time 106 miles, across the Alps. So instead of running we hike.

In the twenty-odd years since we last saw each other, he's been on quite a journey. After we left school he spent several years battling multiple addictions, including a dalliance with Class-A drugs and some dealing on the side. That ended when he was beaten bloody by a rival gang of dealers and, mortified by having to reel off a list of all the drugs he'd taken to the hospital doctors, he moved back in with his parents and sorted himself out. Incredibly, through all of this, he kept running. The day after he took that beating he ran his first two-day mountain marathon with a broken jaw, his teeth wired together and under doctor's orders not to exert himself.

Over the next few years he settled into the relative calm of marriage, kids and a working life, first as an outdoor instructor and then as a church pastor. He continued running and represented both England and Great Britain in fifty- and hundred-mile races. Then, when he was at the top of his running game and close to being ordained as a priest, a series of personal set-backs led to a mental breakdown.

Even in the retelling his first decade of adulthood sounds exhausting. According to Marcus, though, his backstory isn't unusual for someone in this sport. 'It's almost a cliché in ultra-running, because so many people come into it with mental health issues,' he says. 'We all want to get away from something.' He's laughing when he says this, but I don't think he's joking. 'If you run for long enough you feel like you've come far enough away from stuff,' he adds.

The psychology of moving forwards through space backs this up. Experiments suggest that literally moving forwards translates into a figurative sense of progress too, and this can have a huge effect on how we feel about ourselves and our lives.

According to two of the fathers of embodied cognition, George Lakoff and Mark Johnson, our understanding of the world and the language we use to describe it are inextricably linked to the geometry of our bodies and the way we move. Successful people are 'on the up', for example, while on a bad day you feel 'down'. When you 'get over' a problem in life, you 'move on'.[10]

In keeping with this, psychologists have found that a person's direction of movement influences what they think about. Moving forwards inspires thoughts about the future, while moving backwards brings back memories of the past.[11] It doesn't even have to be actual, physical, movement – in lab-based experiments where volunteers watched a starscape that appeared to move either forwards or backwards, and even when they were asked to close their eyes and imagine moving in one direction or the other, the mere suggestion of motion was enough to direct the content of their thoughts.

Research also suggests that moving forwards distorts the way we perceive time. Most people (at least in Western culture – it differs in some parts of the world) move along an imaginary timeline where the past is behind our back and the future directly in front of our chests. But experiments suggest that, when we move, this timeline gets stretched and distorted so that the past feels further away. In one impressively low-tech experiment, volunteers were asked to walk from a starting position (a gaffer-tape line on the floor) to another (a black bucket a few metres ahead), and were then probed on how distant past or future events seemed to be.[12]

That the past feels more distant as we physically move forwards is important, because a major risk factor for depression is the tendency to ruminate, getting stuck in a loop of over-analysing things you've said, done or experienced in the past while getting steadily more despondent. Physically moving forwards can help prevent this by making the bad stuff seem further behind you.

This is definitely true for Marcus. 'When you struggle with depression, there's this assumption that when that person says "I can't be bothered to move", that staying put is what they want. But, actually, when you're depressed it's like you're tied to the chair and you want to get away,' he says. 'Running gives you that sense of: I'm here, and I can end up there. Moving forward gives you the strength to know that you *can* move forward.'

Of course, one of the problems with depression is that, when tied to the chair, it can be incredibly difficult to find the motivation to untie yourself and move at all, let alone run. So, for some people at least, medication can provide the impetus to get going in the first place. In a recent study an increase in voluntary movement was found to be a good indicator that an antidepressant was kicking in.[13]

There's also evidence that depressed people walk differently to non-depressed people: more slowly, hardly moving their arms and assuming a slumped posture, with their eyes to the floor.[14] It seems likely that the depression causes the walk rather than the other way around – and yet a change in walking style has been shown to change the contents of thoughts. In experiments when people walked with a high-energy, bouncing gait they were able to remember more happy words from a list of emotionally charged words, while those who were asked to walk slowly and with little bounce remembered more of the negative ones, even when they weren't aware that they were doing a 'depressed'- or 'happy'-style gait.[15]

Interestingly, trail runners don't slog along the ground like a roadrunner, but instead take smaller steps and bounce off the landscape like a rubber ball. Which might explain the big grin on Marcus's face when he finally persuades me to run, downhill, and back to the car park. 'Imagine that the ground is made of hot coals and you don't want to spend too much time on it,' he says. So, I should spring? I ask. 'Yes!' he says, bounding away down the hill in front of me. 'At the end of a hundred miles,' he calls over his shoulder, 'maybe not quite so springy ...'

*

It's a little-known fact that Charles Darwin struggled with both his physical and his mental health throughout his adult life, which his daily constitutionals presumably helped with. According to the written testimony of his son Francis, though, Darwin wasn't exactly bounding along, trail-runner style or synching his footsteps with his heart rate at 120 beats per minute. Instead he would amble around his thinking path, stroking his beard and seemingly in a world of his own:

> As he paced [...] he struck his heavy iron-shod walking-stick against the ground, and its rhythmical click became a familiar sound that spoke of his presence near us.[16]

That doesn't sound much like a man who was springing off hot coals. Even so, the Darwin-style shuffle has some specific, and important, mind-related benefits of its own, which may have helped him to explain life on Earth in a way that no one else ever had. There is now a growing pile of evidence that if we make like Darwin and get musing on the move, it could help all of us to come up with more creative and original ideas.

Creative thinking is one of those skills that our species likes to claim as uniquely ours, but the sad fact is that very few people find that it comes naturally – at least in adulthood, which is when we could probably make the most use of it.

The root of the problem can be found in the brain and the way that, while hosting its body-wide chatroom, it adds its own ten-pence worth in the form of predictions based on previous experience about what is likely to happen next. This process helps speed up decision-making and reduce the likelihood of surprises, while constantly updating its prediction based on what the rest of the body is saying. It's a role that falls largely to the prefrontal cortex – the logical, thinking and impulse-controlling parts of the brain that are housed behind the forehead.

When you get the urge to do a cartwheel in the supermarket, say something inappropriate in a meeting or jump a red light, this is the part of the brain that jumps in and reminds you not to be so stupid. It's a useful feature in all kinds of situations and saves us a lot of time and potential embarrassment, but the downside is that it does the same job for ideas, shutting down thoughts that are a bit out there but which might just work.

These brain regions don't get fully wired into the rest of the brain until early adulthood, which helps to explain both the unbridled creativity of children and the less than stellar impulse control of the average teen. Once it's fully integrated, though, it works as the proverbial 'box', and thinking outside of it becomes much more difficult.

Difficult, but not impossible. Lots of things can temporarily reduce activity in the prefrontal cortex (a state called hypofrontality) – and many of them involve moving.

One thing that works in our favour here is that, whenever you're moving under your own steam at a pace that feels easy, the activity in the prefrontal cortex gets temporarily turned down, perhaps because the brain reallocates blood flow to the circuits involved in moving and navigation and away from 'thinking'. Since the job of the prefrontal cortex is to narrow down the number of thoughts and memories to the most sensible and obvious, turning 'the box' down a little allows the mind to wander without restriction and, potentially, to make new connections without the chatroom facilitator jumping in to veto them before they've fully formed. Reducing this filter allows access to a broader set of options – ideas that you might not otherwise consider.

Another job of the prefrontal cortex is to direct our attention towards a particular goal and hold our attention on that goal while we think about a solution. According to the work of the Dutch social psychologist Ap Dijksterhuis, for certain kinds of problems this kind of goal-directed, conscious, straight-line thinking is actually the worst way to go about making a decision.[17] Conscious thinking makes use of working memory, a kind of mental notepad where we place information while we work with it to come to a conclusion. This is a skill that relies heavily on the prefrontal cortex, and comes with a catch: working memory is limited to around five pieces of information, plus or minus two. Any more than that and we start to lose our thread.

Dijksterhuis argues that when problems have more moving parts than our working memory can handle – like those that Darwin was wrestling with – we will actually do better if we take conscious thinking out of the equation altogether. In this, his 'unconscious thought theory', being distracted from thinking about a problem allows unconscious processes to get stuck into the problem instead. And since this kind of thinking isn't restricted by the number of mental slots in working memory, it can take far more into consideration at any one time. Then, when a solution presents itself, the answer bursts through into consciousness as a moment of insight, or an 'aha' moment.

In experiments, Dijksterhuis asked volunteers to study the details of several different apartments, all with many pros and cons. One group of people were distracted for three minutes before making a final decision, while another group was asked to choose straight away. The people who were distracted for three minutes made better choices than those who had thought directly about the problem.[18]

Not everyone is convinced that unconscious thought is any better than direct thinking, or even that it exists at all – the problem with unconscious thoughts is that the person having them is unaware of their existence, which makes them tricky to measure. But whatever the cause, there is good evidence that a short spell of hypofrontality not only provides some respite from depressive rumination but also boosts creativity by favouring blue-sky thinking over run-of-the-mill solutions. We know this because experiments at Kansas University showed that when activity in the prefrontal cortex was temporarily knocked out using a type of brain stimulation called tDCS, volunteers were able to come up with twice the number of creative suggestions when asked to think of new uses for everyday objects. They also came up with ideas significantly faster when their idea-narrowing prefrontal 'box' was taken out of the equation.[19]

On a visit to Kansas in 2016 I was able to have a go at this experiment for myself. Once Evangelia Chrysikou, the lead researcher, plugged my brain into the tDCS machine, which was, in turn, connected to a 9-volt battery, I felt my attention drift off to the middle distance. Then, when shown a variety of everyday objects, I had no problem coming up with new ideas on how to use them. Obviously, a dartboard could be used to wipe your feet, the metal bits sticking up would be great for scraping the mud off your shoes. And surely everyone would agree that blowing your nose into a velvet drawstring bag would be far more hygienic than a tissue.[20]

But you don't need to be wired up to a 9-volt battery to get this kind of a creative boost. A recent study by researchers at Stanford University confirmed what Darwin stumbled on over a hundred years ago: that walking has a very similar effect.[21] In a series of experiments the researchers also asked people to come up with unusual uses for a variety of familiar objects. Sometimes the volunteers were sitting while doing this; at other times they were walking. In addition, sometimes they were asked to sit and walk indoors and other times outside. The results showed that, compared with sitting, walking increased people's ability to come up with creative uses for ordinary objects by up to 60 per cent. If they walked first and then sat down, the effect of walking spilled over so that they were more creative for a short time afterwards, too. 'Taking a walk before brainstorming should help improve performance,' the researchers concluded.

In this particular study it didn't matter where people did their walking. Walking on a treadmill while looking at a bare wall was just as effective as strolling outdoors. However, there is some evidence to suggest the opposite – that spending time in green space provides an extra boost.[22] Other research suggests that time in nature works as a kind of reset button on our ability to pay attention. Wherever you do your walking, though, getting into the right state of mind is the most important thing, and ambling along at a comfortable pace seems to be the most user-friendly, and effective, way to get there.

All things considered, then, it's troubling that the world's big thinkers spend more time bent over a desk than wandering the hills and valleys pondering new ways to solve our woes. And even among the common or garden average-size thinker, few people these days go walking just for its own sake (17 per cent of people according to a recent survey, and that includes dog walkers, who arguably don't have much of a choice).[23] Meanwhile a group of economists have sounded the alarm that creative ideas seem to be getting thinner on the ground as the years go by. Coincidence? Perhaps. But it does seem to be a problem worth looking at. In a paper for the US-based non-profit organisation the National Bureau of Economic Research they pointed out that, despite the fact that research effort has been increasing year-on-year for decades, the output from that same research is diminishing.[24]

Even children, who are the most naturally creative beings

on the planet, partly because of their small and not yet perfectly formed prefrontal cortex, seem to be losing their edge. In 2011 the psychologist Kyung Hee Kim, from the College of William & Mary, in Williamsburg, Virginia, compared scores on a standardised test of creativity from the 1990s to the 2000s. Shockingly, she found that the scores had significantly decreased in that period, particularly among younger children. A more recent update of that research suggests that the trend has been getting worse since then. Kim largely blames this on the modern educational obsession with testing. Even so, given the findings that movement can enhance a creative state of mind, and given that it's easier to change individual behaviour than educational policy, she recognises that modern lifestyles also play a part.

'The rise in sedentary lifestyles is a factor in the decline of creative thinking,' she told me in an email, pointing out that the rise of passive play, such as watching TV and other screen-based games at the expense of active play, is a serious problem both at home and at school.

In her view it doesn't necessarily matter what kind of activity you do, whether it's walking, running or physically acting out stories: moving around can help bring ideas forward in a way that sitting around never will. 'Creative thinking is stimulated by physical activity, whether walking, running or active playing,' she says.

One way to help stop the rot, then, is for anyone who is able to do so to get on their feet whenever possible and move forwards at whatever speed feels easy to them. If physically walking isn't possible, or if cycling or canoeing is more your thing, then moving forwards in other ways will have at least some of the benefits as long as you do it at a level that is easy enough for you to be able to forget that you are moving and let your mind wander. Ideally this would be done alone somewhere familiar, so that your thinking mind can switch off, drift away and come back with a shiny new idea. It really is that simple.

Well, sort of. The catch is that the quality of the ideas that come out during a period of hypofrontal musing very much depends on what's in there in the first place, which in turn depends on the experience and memories of the person doing the musing. Memories are stored in widely distributed networks across the brain (and, some claim, the body), which is why one thought can instantly bring to mind something else, like a domino effect. Each person's network is different, because of their completely different life experience. The upside of this, say the Stanford researchers, is that, so long as they can temporarily switch off their filter, each person can tap into their unique network of knowledge and memories for inspiration. And when the 'aha' moment comes, things that may seem totally disconnected suddenly fit together in ways that are so blindingly obvious you can't believe no one else has thought of them. And they may not have done, because they're not you.

There's no shortage of problems in the world needing creative solutions: climate change, famine, war, global pandemics, ageing, population bottlenecks, dwindling resources, you name it. There's plenty for humanity to get its teeth into.

The point here is that the next generation of Darwins will struggle to plumb the depths of their own minds if they spend most of their time sitting around, staring at the nearest screen. Add that to the mood-boosting effects of synching your steps to your heart rate, the memory-protecting power of bone-derived hormones and the mental benefits of physically moving forwards through space, and suddenly sitting still seems like the worst possible way to go about the business of thinking.

How to move: on two feet

- **Time it right:** Walking at the fairly brisk pace of 120 steps per minute (two steps per second) synchronises your footsteps with the heartbeat, adding a small but significant boost of blood to the brain that might contribute to the feel-good factor of exercise.
- **Go somewhere:** Mentally moving forwards through space has been shown in psychological studies to direct thoughts to the future and away from depressive rumination, while making the past feel more distant. Whether on two feet, two wheels or some other means, get out and move.
- **Wander to think:** Walking, or running, at a pace that feels easy turns the 'thinking' brain down and lets the mind wander in a way that boosts creativity and problem-solving. Do it before a meeting for a mental boost.
- **Defy gravity:** Putting weight on your bones stimulates the release of osteocalcin, which improves memory and may future-proof the brain for old age. Maybe add a backpack for extra weight (and snacks).

3

Fighting Fit

Physical strength is an expression of the
total functioning of the organism.
Jean Barrett Holloway

Terry Kvasnik had been training for this moment his entire life. Starting with gymnastics at the age of three and then moving on to breakdancing, martial arts and parkour, he spent his twenties and thirties living the dream, working first as an acrobat in a London West End show and later for Cirque du Soleil. When a car pulled in front of his moped as he was riding along at 40 mph, it could all have been over in a flash. Thankfully, he knew exactly what to do. Or at least his body did.

'It was like my body said, "I've got this, step out of the way, Terry",' he says. 'I just knew ... "I'm going to flip".' And so, he did. Using the moped as a launch pad he dived up and over the car, rolled over on his back and popped back up onto his feet, 10 metres from the wreck of his moped. It was only then that his mind caught up. 'I turned around and sat down again, like, holy shit, what just happened?'

He walked away from the accident with mild concussion, a torn cartilage in his chest and a banged-up knee, but it's a miracle he survived it at all. Ironically, he had only moved to

55

Los Angeles, where the accident happened, to try and find work as a stuntman. 'I often think it was my subconscious trying to get itself a stunt fix,' he laughs.

Kvasnik is, it is fair to say, a fairly expert mover. But whether you're a car-leaping acrobat or ordinary mortal, there is a lot to be said for having life-saving strength and agility in your back pocket for when you need it. Evidence from psychology studies has been accumulating for some time that having the physical skills to get out of sticky situations makes a big difference to how mentally capable and emotionally resilient we feel as we battle our way through life. Becoming a master of your body, in other words, might help you become a master of your mind.

As far back as 1988, a study of teenage girls showed that weight-training that boosted their physical strength by 40 per cent over twelve weeks increased the girls' feelings of confidence about their 'general effectiveness in life'. It also improved their ability to resolve social conflicts that had nothing to do with physical confrontation. The study's lead author, weightlifting enthusiast Jean Barrett Holloway, lamented 'a population of women who exhibit strength levels below their own potential' and were missing out on the mental and emotional side-benefits as a result.[1]

In the thirty or so years since then, women have actually begun to catch men up in the strength stakes, but this piece of good news is tainted by the fact that men are getting progressively weaker. A study from 2016 compared the grip strength of twenty- to thirty-five-year-old students in 1985 with a similar number of modern men. The men of the 1980s could exert 117 lb. of force, compared to millennial man's measly 98 lb.[2]

The next generation are seemingly even worse. A recent assessment of British schoolchildren found that ten-year-olds are significantly weaker than they used to be – showing a 20 per cent decrease in muscle strength and a 30 per cent decrease in muscle endurance since 1998.[3] Worse, their weakness is accelerating with every year that passes, and the trend has been speeding up since 2008. Unsurprisingly, the main culprit is too much sitting and not nearly enough weight-bearing exercise. There are similar trends across Europe and the US.

This is a worry because strength is good for all kinds of reasons. For a start, it has been linked to a longer, healthier life. Studies that followed people over decades found that muscle weakness is linked to a greater chance of dying from any cause, regardless of whether you are carrying too much fat and independent of the amount of cardio you may do.

There is also a link between bodily strength and a healthy brain. A ten-year study of twins showed that greater strength in middle age is linked not only to more grey matter but also to a better functioning memory and a quicker brain a decade later, while grip strength (an overall indicator of muscle power) is associated with a healthier hippocampus.

More important, perhaps, is how physical strength makes you *feel*. Since Barrett Holloway's initial studies, strength training has been shown pretty conclusively to make life feel significantly more manageable, boosting self-esteem and helping people feel capable of meeting physical and emotional challenges.[4]

A brief dip back into the philosophy of consciousness suggests a possible explanation. According to the neuroscientist and philosopher Antonio Damasio, our sense of self – the feeling that there is one 'me' living this life in this body in this moment – is built firmly on our body's implicit assessment of what it can handle.

This is because, as we've already seen, the tissues of our bodies never, ever shut up. They constantly prattle on, like children in the back of a car, back and forth to each other and to and from the brain, commenting on how things are going on the inside. Part of the power of movement is that it allows us to change this commentary in the moment, which has an immediate impact on how we think and feel. But its effects can run even deeper than that. Any movement that strengthens muscles and bones can change the content of the commentary for the long haul. Moving in a way that makes us stronger can dramatically change our sense of who we are and what we can achieve in life.

This assessment comes from the moment-to-moment tweaking of countless physiological dials that keep our biology within a safe, liveable range. This system, homeostasis, comes in three basic flavours: hormones released in the bloodstream; nerve signals to and from our organs; and physical feedback from our muscles, bones and other tissues.

Part of this system is what Damasio calls the body's 'musculoskeletal division', which has the job of updating the brain about the state of the muscles, bones and other parts of the body that are involved in movement. 'Even when no active movement is being performed, the brain is also being informed of the state of its musculoskeletal apparatus,' writes Damasio.[5]

If our eyes are our window on the world, in other words, then our flesh and bones are the vehicle that allows us to act on that information in ways that tip the odds of survival in our favour. At the risk of getting my metaphors in a twist, this is not a passive vehicle, driven by an all-knowing mind, but a chatty one, like KITT, the talking car in the classic 1980s TV show (and soon-to-be film) *Knight Rider*, who constantly witters on about the odds of mission success. It's not surprising, then, that the way we feel has a lot to do with whether our particular vehicle is creaking and rusty or primed and ready to turbo-boost over the next road-block.

If we let our bodies become weak, the message coming from the musculoskeletal division of the self will read: stiff, feeble, could definitely do better. And if, as psychologist Louise Barrett puts it, this read-out feeds directly into our perception of 'what that body can achieve in the world', then it's hardly surprising that sedentary lifestyles have been linked to anxiety and low self-esteem.[6]

The good news is that we can upgrade our vehicle at any time. Adding capacity to the muscles, bones and other weight-bearing tissues of the body is not only expressed internally as a tangible feeling of being capable in all walks of life: it also shows on the outside, giving off clear messages of confidence in posture and behaviour. As if to prove that the mind–body loop never ends, this change in posture then feeds back into our mental state.

Micah Allen, a neuroscientist at Aarhus University in Denmark who studies interoception – how the messages coming from the body feed into our inner lives – says that, in his own experience, increasing his strength through climbing had unexpected side-effects that spilled over into life and work. 'Climbing is something where you start off and you literally don't know if you're going to have the strength to make it to the top of the route and back down,' he says. But as he progressed, he started to notice a creeping sense that he was more capable in other walks of life too. 'Before, if I had a meeting with someone, I could be more easily intimidated or more nervous,' he says. 'But something about that implicit sense of confidence of knowing what my body was doing, purely anecdotally to me, it definitely did have an effect.'

There is some evidence to suggest that his hunch is correct. Research confirms that people who do more physical activity tend to score higher on a scale of 'global self-efficacy' – which measures their sense of how much control they have over their lives. The effect is seen in healthy adults, children and adolescents alike, and according to studies comparing different forms of exercise, strength training has faster and more powerful effects on self-esteem than improvements in cardiovascular fitness and other kinds of exercise that focus on balance or flexibility.

Feeling capable and in control is, of course, the polar opposite of feeling anxious. It's a common misconception that anxiety is about living in a state of abject terror. Often (and I am speaking from experience here) it is more of a rumbling undercurrent of uncertainty about life and whether you can cope with the challenges it holds. Studies using weight-bearing exercise as a treatment for anxiety have found that getting stronger seems to make at least some of that angst go away, boosting self-worth while reducing symptoms of anxiety and improving sleep.

Similarly in depression, the overriding emotion isn't necessarily sadness, but more a pervasive, visceral feeling of 'I. Just. Can't'. Studies consistently show that weight training seems, literally, to lighten the load. Perhaps strength training helps to change the internal feedback from 'nope' to 'let's give it a shot ...', providing a sense of confidence that the body can deal with the trials of life, allowing the 'thinking' mind to take a break.

This raises the important question of whether there is a link between rising levels of anxiety and depression in our society and an increasing proportion of people who are physically weak. Scientifically, no one has looked at this question in detail, so it's difficult to be sure. However, given the evidence that sedentary lifestyles lead to anxiety and that strength training improves both self-esteem and symptoms of mental illness, it seems both likely and ripe for investigation. It's entirely possible that we, in Western society, have spent the past few decades turning ourselves into cosseted caged animals who no longer have implicit faith that our bodies are up to life's challenges. In short, poor mental health might be part of the price we pay for a cushy life of sofas and supermarkets.

The easy life may also be making large numbers of people feel down in ways that are less dramatic than depression but which can still colour life in shades of grey. According to Damasio, the unconscious messages coming from the body provide not only the basis for the self but also a kind of undercurrent to our consciousness that sets the mood for everything else that happens.[7] These 'background feelings', as he calls them, act a bit like the soundtrack of a film: they have the power to make us feel happy, sad, hopeful or on edge, for reasons that we can't quite put our finger on.

It stands to reason, then, that if we can change the tune on our background feelings we can also change the way we feel. Perhaps if we can make our bodies physically stronger we can change that background music, from the sinister discords of a psychological thriller to the rousing harmonies of a superhero theme.

Get strong for what?

I'm not sure what music is playing in the background of Jerome Rattoni's life, but I'd be willing to bet that it's both powerful and upbeat.

I'm in a tiny gym, tucked under the railway arches in Hackney, east London, with twenty or so fitness instructors, all of whom are looking at Rattoni in open-mouthed admiration. He has just hopped up and grabbed a bar that's more than a foot above his head and then, with no visible effort, pulled up his entire body weight until the bar is by his waist. Finally, he hops up onto the bar and crouches on top, elbows rested casually on his knees as he grins down at us.

'What is the point of a pull-up?' he asks. The instructors and I start to mumble about upper body strength.

'No,' he says, dropping back to join us on the floor. 'The point of a pull-up is to get up on top of something. Otherwise why would I bother? I go up, I come down again. I could have just stayed down.'

His deadpan delivery and French accent, combined with a Gallic shrug and a twinkle in his eye, not only make an excellent point about the futility of what most people do in the gym but also hint at a way that we might make strength training more effective, both physically and mentally, and a whole lot more fun.

Rattoni is a master instructor for MovNat, a system of fitness training that emphasises natural human ways of moving. Devised in 2008 by another Frenchman, Erwan Le Corre, MovNat is like a lovechild of forest bathing and parkour, which focuses on getting strong in the natural environment by climbing, jumping, balancing, swimming, running, lifting and carrying the way our ancestors presumably did. In this brave new world, true fitness is not about lifting weights to get big muscles or running to score a personal best: it is about having a body that is strong and agile enough to move like the animals that we often forget we are. Once we have these skills, the theory goes, we will be free to stride confidently through the world, sweeping aside danger, leaping over obstacles and laughing in the face of stress.

It sounds good. And there's no doubt about it: if a tiger were to appear in Hackney Park, where, to the amusement of the local dogs, Rattoni has brought us to practise our crawling skills, he would be the only one getting out of here alive. Over an intensive weekend of natural movement training he demonstrates how to swing through trees, shoulder heavy loads and run for dear life while leaping over obstacles. 'Vaulting,' he says with a totally straight face, is just 'running with something in the way.'

Le Corre took his inspiration from the work of Georges Hébert, a French naval commander from the early 1900s who, impressed by the strength and agility of the indigenous hunters he saw on his travels, introduced a new form of physical training to the naval recruits, based on perfecting the full range of natural human movements. Hébert's work was lost in the aftermath of the First World War and was forgotten for decades but has recently been resurrected thanks to natural movement enthusiasts like Le Corre and early adopters of parkour, the urban equivalent.

For Hébert, being strong was about more than just brute strength: it was a moral responsibility to have the kind of body that you could use to spring into action in an emergency. His mantra was 'be strong to be useful': if we can't run, climb, jump or swim to safety and we can't throw with decent aim, then we are not truly capable of looking after ourselves or, indeed, anyone else.

Terry Kvasnik couldn't agree more. When he's not dramatically leaping over cars, he spends much of his time teaching children to master gravity-defying flips and acrobatic tricks in his home town of Ojai, California. But while the kids are mostly there for the thrills, he also tries to introduce them to the idea that a strong, agile body is a useful tool in an emergency. 'I don't want to scare them but it's kind of real in California. We've got earthquakes! Yes, we're doing cool tricks but actually you're learning how to utilise your body to adapt to any environment and to be able to escape or evade or help. That, for me, cuts to the core of what it is all about.'

There hasn't been a great deal of research done comparing natural, life-saving movement regimes with other ways of getting strong in terms of what it does for the mind. What little has been done has concentrated on improvements in working memory – interestingly, a capacity that suffers in both anxiety and depression.[8] Nevertheless, if getting strong is a short-cut to telling your mind that you have the skills necessary to survive, then there are worse ways of doing that than learning to move like a wild human should.

And anyway, it's a lot more fun than going to the gym. Granted, you do feel a bit silly crawling around public parks, but skills like crawling work on many muscle groups in one go, making it potentially more effective than contorting into awkward positions to crunch a few muscles at a time. After a morning on all fours I ached in corners of my body that I didn't know contained abs. I put it to Rattoni that more people would take up crawling if it was rebranded as something like a 'loco-plank'. 'No,' he says, batting the idea away with his hand. 'People just need to get over it and crawl.'

The rise of things like MovNat, as well as climbing and mud-caked obstacle course races, are perhaps signs that people are beginning to fall out of love with exercise the old-fashioned way and are more interested in getting strong as nature intended. In the UK the huge rise in wild swimming – swimming outdoors in lakes, rivers and the sea – is another sign of our appetite for natural movement. The craze began with the surprise bestseller *Waterlog: A Swimmer's Journey through Britain* (1999), by the environmentalist Roger Deakin, and has since persuaded around half a million Brits that getting wet and chilly outside is a sure-fire route to happiness. There's no shortage of anecdotal reports of improvements to mental health, and there are a handful of intriguing studies into how cold water might reset the stress system into a place of calm.[9] There are also provisional study results that suggest that exposure to cold water releases a 'cold shock protein' into the blood, which protects the brain, potentially slowing the progression of dementia.

Whatever the mechanisms, there's no doubt that wild swimming, and other forms of natural exercise, are as good as, if not better than, the gym for getting fit and building a six-pack. Rattoni isn't muscle-bound, but he is undoubtedly in excellent physical shape. He says people often don't believe that he got this way from natural movement alone. 'People think you are just out there hugging the trees,' he laughs. 'But MovNat can kick your ass if you put your heart into it.'

You got the power

Whether you achieve a strong body via swimming, carrying rocks or climbing on top of them, the obvious question is why physical strength should translate into mental resilience. Nailing down the exact mechanisms is tricky, partly because it's not easy to do experiments in mice and rats that are the human equivalent of weightlifting. Rodent exercise tends to involve wheel-running, which means that it's difficult to separate aerobic exercise from strength alone. Then there's the fact that rats and mice don't have the same kind of minds as we do, so it would be tricky to know whether they gained mental health benefits in the same way as us.

Even so, research in humans hints that it isn't necessarily the process of building big biceps and a six-pack that confers the mental benefits. A recent analysis of nearly fifty studies on mental health and physical strength found that the improvements in symptoms of anxiety and depression follow weight-training regardless of significant changes in the size of the muscles.[10]

At first, this sounds like a blow to the theory that physical strength makes you stronger on the inside. In fact, it is actually explained by a simple fact of life. Without wanting to sound too much like a motivational poster: you are already stronger than you think.

This is because our muscles almost never use 100 per cent of the force that is potentially available to them. Some is always left as a back-up in case we have miscalculated the stresses that are about to be put on the muscle. Because of this, the first thing that happens when we take up weightlifting has nothing to do with adding more protein fibres to the muscle. Instead the body, rather sensibly, releases spare capacity that it already has. Once you get used to lifting heavier weights – or even just holding up your own body weight – this natural brake is released a little, unleashing a bit of latent power. Meanwhile motor neurons, which connect muscles and the brain via the spinal cord, start to branch out within the muscle, connecting to more fibres at once and consequently generating more force for each contraction. Along with gains that come from improving technique, these early changes can be beneficial on the inside, regardless of whether anyone can see them on the outside.

Tiny dancer

Then there's the fact that dumb muscle is far from the only component of strength. The kind of power that sent Jerome Rattoni springing up for a pull-up, that helps ballet dancers leap and twirl and that allows ninjas to land silently has a lot to do with the connective tissue. This includes the tendons, which link muscle to bone and fascia, a strong yet elastic type of tissue that is woven through and around the muscles.

Tendons translate muscle contractions into action and have elastic properties that add extra power when needed. Jumping and springing are two examples. Kangaroos and gazelles achieve their bounding skills not through the muscles of their, frankly, skinny legs but thanks to the elastic qualities of their connective tissue, especially their spring-like tendons. There has been less research on muscle-related fascia but there is some evidence that it, too, can transmit force across the body from one region to another, potentially adding extra explosive power.

The elasticity of the connective tissue in our shoulders is also what makes humans the best throwers in the animal world. Pulling an arm back overhead pulls the ligaments and tendons tight, like a rubber band. Twisting at the waist and cocking the wrist puts more power behind it until – bam – it's released.[11] Our shoulder design not only allowed us to become better hunters, bringing down far bigger and stronger creatures than ourselves, but also, some say, contributed to our ability to think ahead, because hunting by throwing rocks and spears required not only brute force but also an ability to predict where the prey will be by the time the projectile meets its mark. And, in my experience, throwing a ball or stick as hard as possible is a great way to release anger or stress, especially if, for added satisfaction, you aim for a target.

The most impressive demonstration of explosive fascia power that I have ever seen was at a Kylie Minogue concert in 2002. It was during a rendition of 'Confide in Me', in which Kylie was playing the role of a sexy but concerned police officer and our friend Terry Kvasnik – in his stage acrobat days – was playing a troubled young hoodlum with fabulous abs. Having flipped around the stage, walked down two flights of stairs on his hands and done some impressive martial arts moves in Kylie's general direction, he crouched in front of her hugging his knees to his chest. As the music rose to a climax, his gaze slowly moved up to Kylie's face and then, in one explosive movement, he leapt into a back somersault and landed back in a crouch, a foot or so further away from Kylie. She stepped forward and the cycle began again. Crouch, somersault, crouch, four times, without so much as a wobble.[12] Afterwards, discussing the show with a friend, it was all we could talk about. 'Did you see that guy? How is that even possible?'

Eighteen years later I tracked British-born Kvasnik down to California to find out how he did it. He tells me that the inspiration for this particular flip came from a fascination with the explosive power of Kung Fu. A chance meeting with a group of Shaolin monks, who were touring the UK and happened to be using the same London rehearsal space, gave him an opportunity to learn from the experts.

'It's almost a cat-like strength,' he says. 'A cat can be sleeping and look like it's dead. You've only got to touch it and, boom! It flips to the ceiling!' The key to the Kylie flip, he tells me, was to relax into the crouch, mentally and physically coil the spring and then release. In martial arts this relaxed power is called 'song' (pronounced 'soong'), the strong and supple power of a jaguar as opposed to the brute force of a thrashing crocodile.

Past that, it was a matter of practising these explosive movements over and over, building up to the next level. 'Crouch, jump and up tuck high … If I can jump that high, and I can do a backflip, then in theory I can do it.' Then it's a case of practising on a soft surface, then a hard one and finally in front of one of the world's biggest pop stars and an audience of thousands, trying your best not to knock her teeth out.

He still performs – and can still flip at the drop of a hat – but now dedicates much of his time to teaching children how to harness this kind of explosive power. In the process of tapping into a superpower that they don't realise they have,

Kvasnik believes that they develop an ability to access hidden mental and emotional reserves that will last a lifetime. The kind of resilience and confidence that leaves them ready to tackle anything.

The key to accessing all this is to mentally tune in with the body – boosting the kids' powers of interoception. Our sedentary lives often leave us up in our heads and with little awareness of what our body is doing, he says, so the foundation of this approach is to show the kids how to mentally tune in to their body. Kvasnik starts each class with breathing exercises and practice at becoming consciously aware of how various body parts feel right now. Then come stretches, also with an emphasis on being consciously aware of how the muscles feel when sitting or standing the right way up. Then, and only then, can you begin to unleash that feeling to go upside down, sideways or anywhere else you might want to go.

'It's just getting them to say, "OK, I can feel my legs right now" ... So when, later, I say, put your mind in your legs as they're going over your head, they've got an experience that they can draw on.' Sometimes, he says, the kids struggle to get the idea. But when they do, the results are explosive. 'When they realise that they can use this power like an accelerator, but only if they put their mind in it ... when that breaks through, it's just like night and day,' he says. 'If you really work it, it unlocks something incredible.'

It sounds amazing, but the bad news for the rest of us is that ageing, a lack of movement, or both, tends to make connective tissue stiff and inflexible, whether your mind is in it or not. This makes cat-like agility more difficult as the years tick by. According to fascia researcher Robert Schleip, of Ulm

University in Germany, inflexibility sets in because over time the fibres that make up the fascia become tangled, sticky and less elastic. Under the microscope they start to resemble a tangled ball of wool rather than a neat, stretchy net.[13]

Exercise, and movement in general, keeps it springy and strong. Schleip speculates that specific fascia-targeted training, such as practising explosive jumps with soft landings and working on releasing power through a joint's full range, could help bring gazelle-like springiness to anyone who cares to put in the time and effort. So far, studies investigating whether this significantly increases strength or reverses age-related stiffening have been inconclusive – at least in mere mortals.[14] But Kvasnik's experience suggests that, with hard work and enough time, explosive power might be more achievable than you think.

How to keep fascia in good working order is something I'll come back to in detail in Chapter 6. For now, the important thing to remember is that, while beefing up muscles may help you feel powerful, it is by no means absolutely necessary. We all have a reservoir of strength that we don't know that we have, and tapping into that strength doesn't need to involve morphing into Arnold Schwarzenegger. Far more important is to keep moving and stay as strong and springy as humanly possible. That way, the message that our tissues are sending the rest of the nervous system will be more likely to say – relax, everything is under control.

Fighting back

This sense of visceral safety is never more important than after experiencing trauma. But because of the way that trauma affects the body and mind, finding this sense of peace can be incredibly difficult. But here too there are signs that movement can help.

Complex post-traumatic stress disorder (PTSD) sufferer and blogger Sonia Lena has written particularly eloquently about how the martial art Krav Maga helps her to deal with flashbacks and feel more in control. 'I can't tell you what it is exactly about this discipline that pulls me back from the edge,' she writes.[15] 'All I know is that at first, when my instructor's hands went around my throat in a chokehold, I would be flung straight into the claws of a panic attack. Now … well, not so much.' Her theory is that her body has internalised the knowledge that she can physically defend herself, which has loosened trauma's hold over her mind.

An event, or series of events, counts as traumatic if it exceeds a person's ability to return to a baseline of calm afterwards.[16] This could be a life-or-death emergency that it's impossible to escape – being trapped in a collapsed building, for example, or pinned down by an attacker – or it could be the not necessarily life-threatening but socially and emotionally painful experience of being repeatedly criticised or humiliated by someone who has power over you.

In these situations our hard-wired alarm system well and truly goes off, sending our heart rate and blood pressure through the roof and tensing the muscles like a coiled spring. In an ideal world, we would then use this rush of energy to get out of danger, throw the bully across the room or, at the very least, run away after having dealt a crushing put-down. Then, once the threat has passed, we can finally calm down and get back to baseline.

In trauma, though, this neat sequence of events doesn't

happen, and some trauma researchers believe that this lack of enactment explains why survivors so often find themselves catapulted back into reliving the event, seemingly unable to make sense of what happened and unable to move on. As well as flashbacks, another common response to traumatic experiences, particularly if there was no escape route or a person was overpowered, is for the victim to collapse mentally and emotionally, checking out of the situation altogether. This can lead to dissociation, an eerie feeling of observing your own life as if looking through a window and in people who have been emotionally abused, a tendency to avoid eye contact in social situations.

This has led some trauma researchers to wonder if finishing the physical job of fighting back or running away could help finally put the experience to bed, taking people out of the cycle of fear for good.

This idea has been made popular by Bessel van der Kolk, a psychiatrist and trauma specialist at Boston University. In his work, and in his bestselling book *The Body Keeps the Score,* he argues that the reason that talking therapies often don't work for post-traumatic stress is that you can't reason your way out of a full body reaction to danger. For some people, mentally raking over the traumatic event in detail might actually make matters worse, pushing them straight back into fight-or-flight without giving them any new tools to make sense of what happened. Or it might lead them to retreat to a place of emotional numbness where true recovery becomes even more difficult.

Van der Kolk, along with psychiatrists Pat Ogden and Peter Levine, believes that flashbacks and dissociation persist largely because the action of responding to stress wasn't finished. If a person can help the body to finish the job, they say, by learning and practising defensive or escape-based movements, it allows homeostasis to finally run its course and a feeling of safety to return.

That's the theory. And it isn't a totally new idea. It is based on the little-known work by the French psychologist Pierre Janet, who in the early 1900s wrote about the importance of taking successful physical action to prevent the traumatic memory from becoming lodged in both body and mind. In his book *Psychological Healing* (1925) he wrote about the 'pleasure of a completed action' and suggested that people 'suffering from traumatic memory have not been able to perform any of the actions characteristic of the stage of triumph'.[17] This, he argued, meant that they have never had the closure that their body – and mind – has been crying out for.

For anyone who has ever been bullied and wished they'd fought back, it instinctively feels like something that might work. And there is some evidence from animal studies, and a few human ones, that movement is a key part of the physiological process of returning to 'safe' mode.

Studies have shown, for example, that mice and rats only become traumatised by a terrifying ordeal – usually a cage with an electrified floor – if they are also locked in and unable to escape. If, however, after having become traumatised by that experience, they are put in the situation again, but this time allowed to run to safety, the trauma goes away.

Other experiments suggest that any kind of intense physical activity might do the same. Rats who were allowed to fight each other after the same kind of stressful experience recovered faster than those that were put back into their cage to rest, suggesting intense muscle activity acts as a signal to the nervous system to turn off fight-or-flight mode.

There's some evidence that this might be true in people. As with depression and anxiety, there is pretty good evidence that intensive exercise can reduce symptoms of post-traumatic stress. Military veterans with post-traumatic stress were recently found to benefit more from therapy that has a physical exercise element, and a review of several studies recently concluded that yoga and resistance training also reduce symptoms of post-traumatic stress.[18] What's less clear is whether specific kinds of movements during therapy, such as fighting, shoving or otherwise defending yourself, are more effective than cycling up a hill or doing squat thrusts to exhaustion.

As Sonia Lena's case suggests, anecdotally, fighting back can be useful. And a handful of small pilot studies have shown that body-oriented therapy with movement practices (such as learning an assertive 'stop' movement) has shown promise in other people with complex post-traumatic stress. This is a condition that can develop after multiple stressful experiences through life, rather than just one traumatic event, and which is particularly resistant to standard treatments such as cognitive behavioural therapy. After body-based therapy, though, people with complex PTSD had a significant reduction in depression scores and an increased ability to cope in work and social situations. Two people had reductions large enough to be no longer considered as suffering from post-traumatic stress.[19] Bessel van der Kolk has done a study of yoga as an add-on therapy in PTSD and found something similar. After a ten-week course just over half of the yoga group no longer met the criteria for PTSD, significantly

Move!

more than the 21 per cent of control participants who took part in a trauma support group.

Pat Ogden says she is working on a study to answer the question of whether it's exercise per se that helps or specific fighting moves. What she could tell me was that they have plenty of 'anecdotal reports from clients and students who say learning sensorimotor psychotherapy has transformed their practice and offered new hope and positive outcomes'. And while the evidence may not be there just yet, she says, 'people are realising the importance of the body in trauma treatment'. There has certainly been a rise in body-based trauma therapies, from Ogden and Levine's sensorimotor psychotherapy to those based on boxing, yoga and martial arts.

It may turn out that exercise in general, teamed with talking therapy and specific movements, are all required for successful trauma recovery: learning and rehearsing specific movements as part of therapy may add a sense of mastery and control, with an expanded repertoire of physical tools to call on should the need arise. Together, this could help a traumatised person, in the words of the psychiatrist John Ratey, to 'actively learn a new reality'.[20] Whatever this new kind of embodied therapy ends up looking like, just sitting and talking will surely have to become a thing of the past. Physical mastery of movement is too important to leave out of the equation.

Breaking the cycle

There's one final reason why working on physical strength is so useful for overcoming trauma. It's the surprising finding

76

that emotional upheaval can leave a person with not only emotional scars but physical weakness too.

Studies with traumatised first responders who attended the World Trade Center site in 2001 found that they had almost half the grip strength of age-matched controls a decade later.[21] A separate study found that another group of first responders – who were generally fit and healthy at the time of the 9/11 attacks and probably, given their profession, stronger than average – were significantly more likely to have movement-related problems a decade later, including slower walking speed and greater difficulty getting up from a chair.[22]

If trauma reduces strength and getting stronger can help recovery, it makes sense that strength training could help people get back to full mental and physical health. There's also an argument that it could prevent trauma or stress from taking too deep a hold in the first place, by helping people bounce back soon after the event rather than years later. Giving these skills to young people, particularly those who are growing up in poverty or with social disadvantages, could go a long way to stopping mental health problems from taking root.

Dale Youth boxing club has been putting this idea into practice for years. The gym, in Ladbroke Grove, west London, has a reputation as one of the best in the country. Since it opened more than thirty years ago it has produced more than a hundred amateur champions, one Olympic gold medallist and two super-middleweight world champions. Squeezed in under the roaring A40 Westway, a major artery into London, the club is a community enterprise, funded by donations and staffed by volunteers, many of whom trained here as kids. Its charitable status allows it to charge just £1 per session, to stay true to its aim of keeping inner-city kids off street corners and busy working on their strength, resilience and presence.

A few years ago the gym became known outside boxing circles, for tragic reasons. From 1999 to 2017 it was housed on the first floor of Grenfell Tower – a 1970s council-run tower block that, in 2015, was made less of an eyesore by being covered with smart new aluminium panels. Unbeknown to the more than 300 residents, the cladding was as flammable as it was shiny. When fire broke out in one of the flats on 14 June 2017, flames tore through the cladding, killing seventy-two people and leaving more than 250 others homeless. The boxing club was destroyed too, and the kids – some of whom had lost friends in the fire – were left with nowhere to go to let off steam – or to work through their emotions.

In the years since the fire, the community has worked hard to make the club bigger and better than before. It hasn't been easy: this is one of the most deprived areas of London, and it butts up against one of the richest. It's not easy to keep kids on the straight and narrow when the straight and narrow is so staggeringly unfair.

But that's where boxing comes in, Dale Youth coach Moe Elkhamlichi tells me. 'If you look at kids nowadays out on the streets, leaving school at sixteen years old, no real opportunities, no real guidance, their self-belief is not there, their confidence is not there because they've probably never been told that they could be great,' he says. 'What boxing gives more than anything is self-confidence, self-belief.'

In Moe's experience – he has been a trainer for twenty

years – this self-belief extends outside the boxing ring and into everyday life. As we shout to each other over the cacophony of gloves on punchbags, he points out two kids in the gym whose school grades and behaviour have vastly improved since they started training here and another whose angry outbursts at home have all but stopped. Part of this comes from the discipline they get from the coaches. 'We run a tight ship here,' says Moe, and the kids know that they are expected to train hard. But when that training starts paying off, the feeling of being able to hold their own leaves the gym with them and hopefully follows them into adulthood. 'If they can walk into a room crowded with people and feel confident, if they can go into an interview and feel confident – if we've done that, then we've done our job haven't we?'

How to move: for strength and resilience

- **Work your muscles:** Getting physically stronger (with or without an increase in muscle size) reduces anxiety, relieves depression and increases self-esteem. It doesn't have to involve pumping iron: using your own body weight works just as well.
- **Move like a human:** Mastering the moves your body was made for feels great, as does the knowledge that you can run, climb, swim and jump to safety if need be. Forget the gym and learn to move the old-fashioned way.
- **Fight:** Particularly after trauma, learning the physical vocabulary of fighting back can help create an embodied sense of safety. Ideally, do it with a therapist in case difficult issues come up.

- **Connective power:** Don't neglect jumping. Working on your gazelle-like springiness, and learning to land silently, makes for healthier connective tissue, which can feed into an overall sense of physical and mental mastery.

4

Slave to the Rhythm

There is perhaps no stronger behavior to unite humans than coordinated rhythmic movement.
Jessica Phillips-Silver et al., *Music Perception*[1]

Watching Kevin Edward Turner dance, leaping and rolling in battered chinos and a polo shirt, it's hard to imagine him being still for long. He turns his back to his dance partner, they link arms and he rolls effortlessly across her back, his legs kicking up in the air on the way. She is delighted, he grins from ear to ear and the seven other dancers in the room wish we could try it too.

I've come to the north of England to join in with a dance group for young people with mental health issues. We're in one of Manchester's old canal-side cotton mills, which once fuelled the UK's industrial revolution by way of steam power and child labour. Now it has been restored as a calm and welcoming community space, with local artwork hanging from the exposed brickwork, shelves packed with books and trailing plants at the windows overlooking the canal. Tonight the energy in the room comes from Turner, a choreographer and dancer who is passionate about how dance can transform the mind.

This is something he knows from experience. In 2013 he experienced a worsening of the depression he had suffered since his teens, this time with episodes of psychosis that were serious enough for him to be hospitalised under the UK's Mental Health Act. He had been a dancer since the age of eight, and the first sign that all was not well was that he no longer wanted to move. 'There was a period of time when I was very inactive and I found it very difficult to motivate myself,' he says. 'It created a perfect storm for me to get worse.'

Dance provided a lifeline that gradually led him back to health. 'It was a very slow process of getting my strength back – my mind, my soul, my body, however you want to think about it,' he says. 'I know, for 110 per cent, that being able to use movement and dance to express what was going on in me internally was a big part of me getting back to being able to work again and do what I enjoy most.'

According to a growing body of research, Turner is on to something. Dance is emerging as a vital tool to maintain the balance between what's going on in the body and how that feeds into our experience of life: one that is crucial for us to function properly as human beings.

This isn't just about dance making you feel happy. It's far more important than that. Dance and other forms of rhythmic movement plug into specific aspects of our biology in ways that help us understand and regulate our emotions, providing a fundamentally human way to connect both with ourselves and with each other.

If that's the case, the vast majority of us are selling ourselves short. Only 7 per cent of Americans and 6 per cent of British adults dance for fun, a number that has been declining for at least a decade.[2] Meanwhile, our collective mental health is a mess. An epidemic of loneliness is striking down not only the elderly but also young people, who on the face of it are more connected than ever before. In a recent survey, almost 50 per cent of eighteen- to twenty-four-year-olds admitted to feeling emotionally disconnected even when they are constantly surrounded by people in both the real and the virtual world.[3] Depression and anxiety are running rife through every age group, and schoolchildren are increasingly self-harming as an outlet for difficult emotions. Against this backdrop of suffering, if greater emotional literacy could be as simple as remembering to get up and dance now and then, that doesn't seem like such a bad idea.

Born to dance

The question of why humans dance – and why other animals tend not to – has generated a huge amount of debate over the years. Some have speculated that it began as a form of physical storytelling;[4] others insist that it is a way of putting on a display to show members of the opposite sex that you are fit, strong and co-ordinated and have what it takes to survive in the wild.[5] One thing that everyone agrees on is that it has been part of our movement repertoire for a very long time: probably for as long as we have been standing on two feet. The earliest firm evidence comes from 9,000-year-old cave paintings of group dance in India,[6] but we know that humans were making music – and probably dancing to it – far earlier. The oldest musical instruments – flutes whittled from animal bones – date to 45,000 years ago, around the time that modern humans first walked (or perhaps pranced) out of Africa.

Every human culture since then has included movement to music in one form or another, usually as part of a festival or celebration and usually in a group with other people. Its ability to bring people together is so powerful that a handful of religious sects through history have tried to ban it altogether. The film *Footloose* was based on a real, ultra-conservative town in Oklahoma where dancing was outlawed until the early 1980s. Dancing in public is still banned in several countries today, including Saudi Arabia, Iran and Kuwait. Even in comparatively liberal Sweden it is illegal to dance in public except on licensed premises, and in Germany and Switzerland dancing is banned on certain Christian holidays. In Japan the ban on dancing after midnight (originally brought in to reduce post-war promiscuity) was only lifted in 2015.

Whether the authorities like it or not, though, rhythmic movement is very much a part of what makes humans tick. An ability to feel and respond to a beat comes hard-wired at birth. Studies of two- and three-day-old babies have shown that if you play them a regular beat while recording their brain activity using electrodes on their scalps, and then skip a beat unexpectedly, their brains respond in a way that suggests that they notice that something is missing.[7] Just a few months later, this natural affinity for the beat begins to link up with movement. Five-month-old babies already show signs of moving in time to the rhythm of music, a skill that starts to look more like dancing as they begin to gain more voluntary control over their bodies. It is also obvious that, from an early age, moving to music feels good. The same study showed that babies who are better able to move to the beat smile more when they are doing so than those who are less well co-ordinated.

The feel-good factor of dance even works on surly teenagers. I saw this to my own amazement a few years back at a psychology conference for prospective A-level students. It was late in the afternoon, and everyone was hot, fidgety and ready to go home. My job, as the host of the event, was to introduce the speakers, field questions from the audience, ask something sensible if the audience didn't and keep everything running on time. The last speaker of the day was the dance psychologist Peter Lovatt. When Peter, a middle-aged, bespectacled man wearing a wacky shirt and a big smile, bounded onto the stage, you could feel the room contract as 300 sixteen-year-olds sank into their seats. It was a tough crowd, but if Peter noticed or cared, it didn't show. This was a man who went through school preferring ballet to football and left without being able to read.

Incredibly, that same boy grew up to become a research scientist, specialising in how movement helps us think. He is living proof that it can: he used dance to teach himself to read, pretty much from scratch, at the age of twenty-two. Later, I asked him how. The first thing I noticed is how often he would break into song, mid-conversation, particularly when he was trying to think of what to say next.

'OK. Well, um, bum, bum, bum, bum. So, I'll try to work out where to start ... Doo doo doo doo doo doo. So how did it all come about ...?'

This, it transpired, is a trait that he used to help him read. 'The first thing I did was to try to find rhythm and patterns' in what was written down, he told me. It occurred to him in his late teens that he couldn't be as stupid as his teachers said he was, because not only was he able to learn a two-hour dance routine from memory, but he had also memorised the entire lyric to 'Rapper's Delight' by the Sugarhill Gang. So he decided to channel these skills into reading by, first, having a go at reading poetry, which, like rap and dance, has a natural rhythm. He wasn't starting from nowhere: he could read well enough to decipher individual words; the problem was that he struggled to string those words together fluently enough to extract their meaning. Rhythm, he found, helped him to get into the flow. 'The rhythm is a vehicle for getting you through the sets of words to the other side,' he says. It seemed to work. Even today, he says, he loves to read poetry – or anything with a natural rhythm.

Another strategy was to apply a practical skill he used when he got stuck learning a dance – essentially, skipping the hard bit and carrying on. 'If you're learning a dance routine and there's a bit you don't quite know, you kind of mark it, you do something silly with your feet … "Five, six, seven, eight" and back to the routine again. So I did the same for reading,' he says.

Rather than giving up when he hit a tricky word, then, he'd make something up and keep going. In one of his early attempts at reading he tackled a Jeffrey Archer novel which included the word 'icon' – or, by his reading, *ikkon*, a word that made no sense. So he made it up. 'In my head, this *ikkon* had all kinds of forms. Sometimes the *ikkon* would make sense and then it wouldn't make sense. So, I'd have to change what the concept of this *ikkon* was. And it became a mental exercise in improvising and not worrying about certainty.'

He eventually worked out that the '*ikkon*' was a religious symbol, and the story suddenly clicked into place. 'It was about trying not to be defeated by the words I couldn't read like icon or yachts.' Or, ironically, 'rhythm'.

The next ten years were, in Lovatt's words 'a long, turgid struggle' through A-levels (one pass, one fail), a degree in psychology and English ('I never finished any of the books') and a PhD, which he describes as 'like running a marathon with two twisted ankles', perhaps because of the distinct lack of poetry in most scientific papers. Afterwards he got a job at Cambridge University in the Faculty of English – not daring to tell anyone that he failed his English A-level. It's safe to say that his dance-based plan worked. To date, he has written two books and countless scientific papers. His studies have shown, for instance, that learning a structured routine helps people to think analytically for a short while afterwards, while improvisation provides a boost to more creative, open-ended thinking.[8]

It's an impressive turnaround, but the real proof of the transformative power of dance came with those modern-day students. Having started by describing some of his studies, he gently introduced the idea of audience participation, asking us all to stand up and shake out our hands and feet. At this point the cringe factor was palpably high, but he ploughed on, demonstrating a short dance routine that he wanted us to try: '*march, march march, clap, march, march, march, clap.*' Reluctantly, we did as we were told. Then, in between describing more of his studies, he added more moves, including turning on the spot, which raised a murmur of embarrassed laughter as we all bumped into each other. Then he strung on a series of 1970s disco moves: thumbing a lift, the mashed potato and John Travolta's famous arm-pointing move from *Saturday Night Fever*.

The more complicated – and the more ridiculous – it got, the more the audience relaxed and went with the flow. Finally, as a grand finale to the talk, we put the moves to music and the room erupted into life. Just fifteen minutes after taking to the stage Lovatt had transformed a room full of sullen schoolkids into an impromptu disco, the whole room buzzing with energy. Even the teachers joined in, grinning broadly.

It was amazing, and life-affirming. But it also raises the question of why? What is it about a beat that gives us the urge to move, whether via a surreptitious toe tap or a full-on freak-out? And why would such a seemingly frivolous activity – one that involves making enough noise to attract predators while using up precious energy – have evolved in the first place? And why does it feel so good?

One possibility is that it's nothing more than a happy accident that comes as a side-effect of the way that the brain and the rest of the body work together to get us through life with as little damage as possible. Many scientists and philosophers now think of the brain as a prediction machine that is constantly making its best guess of what is about to happen based on what has gone before, which it then uses to guide our actions and behaviour. According to Morten Kringelbach, a neuroscientist at the University of Oxford, the reason we love a regular beat is that it makes it easy to predict what is coming next. When our prediction is right, we get a small hit of dopamine, a brain chemical involved in reward and pleasure.[9]

Because of the way that sound and movement are linked

in the brain, following a beat with our bodies not only feels good but is also pretty much effortless. Brain-imaging studies show that when we hear music, and regardless of whether we move to it or not, regions that deal with planning movements become active at the same time as those involved in sound-processing.[10] These links don't exist specifically for dance – they are there to allow for the kind of automatic, unconscious processing that allows us to catch a ball, or at least duck out of the way, by basing our movements on what our senses are telling us. As we saw in Chapter 1, the whole point of sensory information is to inform our movements in the world.

A beat fires up these same brain–body pathways in such a way that it is difficult not to move in time. It does this via waves of synchronised electrical activity in brain regions involved in sound and movement, so that brainwaves in the two regions start to link up like two pendulums swinging in time. This phenomenon, called entrainment, makes sharing information across the brain easier because a synchronised pulse stands out clearly from the background crackle of electrical information – a bit like a football chant stands out from the background hubbub of a packed stadium. The beat's ability to cut through the neural noise is a key part of the urge to bop along, because it allows us to move to the beat with very little conscious effort.

There is even more satisfaction up for grabs when we give in to this urge and actually move. Timing our movements with the beat gives us a second hit of dopamine, says Edith Van Dyck, a music psychologist at the University of Ghent in Belgium. Plus, she says, it creates a feeling of being 'one' with the music. It might even give us the empowering illusion that we are controlling the beats by stomping them into existence.

My free-form dance experience gave me a particularly satisfying taste of this feeling. This came as part of the 'staccato' section of the evening when, to my relief, wafting around following our hands gave way to foot stamping, arm punching and, eventually, jumping up and down in time to a tub-thumping beat. This, finally, was something that felt like it came naturally to me – the kind of dance that toddlers do before they grow up and realise they are supposed to be embarrassed by it.

If there is one form of dance that you could do anywhere in the world without looking too out of place, this is surely it. Tribal dances from Africa to South America, and from the jungle of Papua New Guinea to the outback of Australia, vary in form and tradition, but at their root they all revolve around making your feet hit the floor and punching your hands in the air while perhaps nodding in time. The same can be said of much of the dance music that has emerged over the past twenty or thirty years.

There's a good reason why this particular kind of dance crosses continents. It comes down to the way the human body is built. As we know, at some point in our evolutionary history our ancestors started spending less time dragging their knuckles on the floor and more time wobbling around on two feet. Once we committed to a bipedal way of life, our bodies adapted to a new mode of movement in which our legs swing like pendulums from our hips. No other animal on Earth moves this way, and it set the stage for us to dance on.

As a Zimbabwean proverb has it, if you can walk, you can dance. That's because all pendulums, even those with a knee inserted half-way down, swing at a regular, predictable rate. A study from 2005 in which people wore motion-trackers while running, cycling and generally going about their business found that this natural frequency varies surprisingly little from person to person. Regardless of a person's height, gender, age or weight, their bodies resonated at a frequency of 2 hertz (Hz), which translates to their heads bobbing up and down twice every second.[11]

The magic number, 2Hz, may have a lot to do with the way we dance. It equates to a rate of 120 beats per minute. By no small coincidence, this is the rate at which the beat falls in almost all Western pop and dance music.[12] It is also the rate at which people are most accurate when asked to tap along to a metronome in the lab. All of humankind is, so to speak, dancing to the beat of the same drum.

As an aside, this may have interesting implications for those who ask why humans seem to be the only species that got the memo about making music and dancing along. Human music is made for humans, by humans, all of whom resonate at 2Hz. In an essay from 2011 the evolutionary biologist Tecumseh Fitch suggested that perhaps other species don't dance to our music because they move to a different rhythm, and while they are deaf to our music, we can't hear theirs either.[13] Having watched my dog (a herding breed) running in circles with others of his kind, co-ordinating as if by some unspoken agreement on direction and speed, it certainly seems possible. But if this is the case, we have yet to tune in to their frequency, let alone learn the steps. Interestingly, some species of animals can learn to follow our beat with enough practice. A few years back, a cockatoo called Snowball became an internet sensation, and a subject of scientific research, for his ability to bob along in time to the Backstreet Boys.[14] Similar behaviours have never been spotted in the wild, however, so we can still consider our dancing skills to be unique for now.

With humans, though, the fact that we are all dancing to the same beat means that it is easy to synchronise not just with the music but with each other too. And this is when the first real-world benefit of dancing comes in. According to research from the University of Oxford, when we move as one, our brains start to lose the distinction between 'us' and 'them'.

The explanation for this is that under normal circumstances we use information from our own bodies, our sense of proprioception, as a guide to what is 'me' and what is not. But when we move in time with others, our brain starts to get confused. Information about our own movements, from the body, becomes blended with the actions of other people, which are coming in via the senses. As a result, the line between self and other becomes blurred.[15] This suggests that dancing together could provide an easy way to tackle loneliness and help us reconnect with the people around us. It could also provide a way of bringing people together who, on the surface of it, have very little in common, or have totally opposite world views. What better way for us to get over our differences than to move together and notice that, from one human to another, we are actually much the same? The historian William H. McNeill described this phenomenon as 'muscular bonding' and argued that it was an important driver of human community, religion and culture through the ages.[16] Central, in fact, to humanity as we know it.

It certainly makes us care more about each other. In experiments, even one-year-olds were more likely to help an adult if they have first been bounced along to music on the adult's knee.[17] Even at this age synchrony makes all the difference to how much you care about others. If the adult bounced them out of time, the toddlers were far less likely to help them out. It sounds harsh, but this tendency seems to follow us through our lives. In adult versions of the same experiments, people were more likely to co-operate in gambling games if they had spent time moving in synchrony first.

Because of this, some scientists are beginning to see dance not as a happy accident but as something that evolved to fulfil an important role in society – to help groups bond emotionally so that they would work together for everyone's benefit.

If this sounds like something that could benefit all of humankind, the good news is that scientists are on to it. Petr Janata, a psychologist at the University of California, Davis, and his team are working on a device that they have nicknamed the 'groove enhancement machine', or GEM: a network of computers and drum pads that allows the researcher to vary the levels of synchrony between volunteers and a computerised partner while they tap along to a rhythmic beat. In the studies, volunteers play a game that tests their willingness to co-operate with one another.

So far, the researchers have only preliminary data, and more work needs to be done before they can say for sure that playing drums together makes people more likely to co-operate. If it does, 'you could imagine taking your GEM to corporate board meetings or meetings of international leaders. Who knows!' says Janata.

Could there be a dark side to all this? The power of synchronised movement is that it bypasses rational thinking altogether and smacks us straight in our emotions. History tells us that in the wrong hands this can become a powerful method of mass mind control. It is perhaps no coincidence that the introduction of the Nazi single-armed salute, which was made compulsory in 1934 and performed several times a day in public and in schools, coincided with the rise in public support for Hitler's ideas.

In his book *Keeping Together in Time* (1995) William H. McNeill argued that performing the salute regularly, and en masse, provided regular 'visceral bonding' opportunities, which sent a powerful emotional message that this was a political movement for the people, and that they were all in it together. The annual mass rallies at Nuremberg, and the marches of anything up to 800 kilometres that youth members of the Nazi Party undertook to get there, did something similar, he argued. Synchronised movement has brought armies together throughout history because it feels good to be part of the group. And when something feels good it's easy to get swept up in the moment and forget to ask whether it is right. The moral of the story is perhaps to be choosy about who you fall into step with.

Get into the groove

The good news is that if you don't happen to know anyone you feel comfortable dancing with – or if the very idea makes you want to run for the nearest sofa and box set – it is perfectly possible to get this effect on your own. According to Petr Janata, it's all a matter of choosing music that gets you 'in the groove' – a term that, in 2012, he rescued from the 1960s and added to the lexicon of neuroscience.[18]

A lifelong musician with wild curly hair, a goatee beard and a love of the Grateful Dead, Janata defines the 'groove' as the experience of hearing music that makes you feel so good it's impossible not to move your body. I meet with him over Skype, where, to my amusement, I find him on a suitably hip orange and brown velour couch, in front of a dark wood-panelled wall.

In his 2012 study Janata played 148 pieces of music, ranging in style from R&B to folk, to a group of student volunteers, then asked them whether it made them want to dance and whether they considered it to be 'groovy'. Despite the range of musical interests in the group, everyone could agree on what that meant. Out of all of them, 'Superstition', by Stevie Wonder, consistently came out on top.

Like many of the songs that scored highly for this quality, 'Superstition' has a syncopated rhythm, which means that a lot of the rhythmic action takes place off the main beat. It's harder to find the beat than a regular stomp, but when we do, we feel like the coolest kid on the dance floor because we've worked out the secret code. And the possibilities for expressing yourself by rolling your hips, stepping to the side and waving your arms to different parts of the rhythm are endless.

This makes us feel good, Janata suggests, because it feels like an invitation to 'join in with the band'. So, even when we are grooving alone, we can still feel connected to something bigger than ourselves. And, as Janata points out, you don't necessarily need to get freaky on the dance floor to get the benefits. 'I'm one of those people who doesn't particularly like to dance, or if I do, it's very, very, very low amplitude sort of movement,' he admits. 'But I'm still totally in the groove. I can have these super-rich experiences even though I'm not busting dance moves.'

Altered states

There may be more to moving to music than bonding with the group, however. Julia Christensen, a dancer-turned-neuroscientist at City, University of London, believes that getting locked into the beat can also catapult us into an altered state of consciousness – one in which we are physically incapable of processing our stresses and woes.

At any one time we are consciously aware of a small portion of what is going on in and around our bodies. We just don't have the bandwidth to consider all of the myriad inputs at the same time, and for good reason: if we could, we'd be constantly struggling to make sense of the sensory overload. Instead we direct our attention to whatever is the most urgent bit of information at that point in time – hunger, an itchy jumper, an urgent email, the rush to get the train, whatever it happens to be. Turning your spotlight of attention to one of these things will temporarily make you forget all about the others.

One explanation for how we pay attention is that brainwaves in various parts of the brain relevant to our current goal start to sync up and beat in time, rising above the background chatter of brain activity.

If that sounds familiar, it should, because it's the same process by which our movements become tied to the beat. And this, Christensen suggests, could explain how music can hijack our attention so easily. Once all the available processing power has been used up on the rich sensory experience of moving to the music and controlling our body movements, any mental processes that involve introspection, fretting about the future or worrying about the past go out of the window. It's an all-consuming experience that acts as a holiday from our inner thoughts and worries.

This is the drug that human society was using before we stumbled on chemical equivalents. It was this, Christensen says, that accounts for the appeal of a trance-like state in both tribal ceremonies and rave culture. The idea is that, when you come out of that state, the feeling of calm, clear-headedness and connection lingers on. And if that's not something that modern society needs, I don't know what is.

Know thyself

The disclaimer about my own dance experience is that I can't be sure that all that stomping and whipping my hair would qualify, to the trained eye, as dance. I certainly wouldn't want to watch a video of it.

According to Kevin Edward Turner, though, that's only because, like most people, I have the wrong idea about what dance actually is. Most of what we see as dance is professionals showing off meticulously crafted routines that have been honed to perfection. It looks great but puts us off having a go. But that, says Turner, is like refusing to play five-a-side football because you can't kick a ball like Cristiano Ronaldo.

'People think, for me to be a dancer I've either got to be able to spin on my head ten times or spin on my feet ten times, and I can't do that, so I'm not a dancer ... Whereas I say, yes you can be a dancer, you can move your body but in your own unique way, expressing your story and your experiences.'

What he's talking about here is not so much about shaking your booty until you go into a trance but using deliberate movements to express and understand your most personal, innermost feelings. It's an idea that sends most people running for the hills, myself included. But there is a mounting body of evidence to suggest that it might be time to get over ourselves.

Scientifically speaking, there is still some debate about what emotions actually are. Some believe that they are a brain-based phenomenon, which then stimulates changes in the body, such as a raised heart rate and sweating. Others believe that the physiological reaction comes first, and that the brain provides context and meaning to the bodily changes – we feel afraid because our heart is pounding, not the other way around.

Whatever the exact order of events, it's pretty well established that emotions are a mind–body phenomenon. And not only are basic emotions expressed very similarly in everyone's bodies but we can read them in other people's movements without any training at all.

This makes sense if you subscribe to the view that, rather than dance being an evolutionary accident, it is actually an ancient form of language that was the precursor to the spoken word. Charles Darwin himself pointed out in 1872 that people – and animals – communicate their emotions to one another through body language that can be read by all other members of their species. Particularly in a social species like ours, being able to communicate our emotions is essential to making our close-knit societies function. And before spoken language came along, we did this, according to Darwin, by way of movements and gestures.

Sure enough, countless experiments have shown that people can reliably read emotions from the movements of people's whole bodies, and even just from seeing parts of someone's body in motion – arm movements while miming drinking from a cup, for example. We can also pick them out from point-light displays, where lights placed on moving joints are recorded and turned into a moving stick-man. Even children as young as five can do this, and you don't need to speak the same language as someone, or even know anything about their culture, to read them like a book. In one study dancers performed classical Hindu dances from the *Nāṭya Śāstra*, a Hindu text dating back at least 2,000 years, which uses specific movements to express nine basic emotions, including anger, fear, disgust, amusement and love. Volunteers from both America and India were equally able to identify the emotions, despite the Americans having never seen this particular style of dance.

Turner says he never has to teach anyone how to express an emotion in dance. 'Some people need to be encouraged because confidence and body issues can be part of it, but if you are able to create a safe, trusting environment then people can really access things, feelings, emotions, physicality that they probably thought they never could,' he says. 'Your body is the perfect way, I believe, for you to be able to work through any issues that you have.'

As if to prove the point, in 2015 he choreographed and performed a dance piece called 'Witness' in which he explored the impact his illness had had on himself and the people around him. He now spends a lot of his time working with young people suffering from depression, anxiety, poor body image and chronic pain.

It takes a certain charisma to encourage vulnerable young women to dance their feelings, but, as I found out when I joined the group, Turner has it in spades. With an infectious enthusiasm, he has the air of the cool, mischievous friend of an older brother, but with a gentle, earnest touch that lets everyone know they're in safe hands. The girls in the group clearly adore him. 'Kevin's great,' one of them tells me. 'He's like a bundle of energy.'

He leads us in a warm-up, urging us to 'surrender' to the physical forces on the body as we move around the room. As if by magic, the whole group subtly changes the way we move, as we tune in to gravity, really feeling our feet on the floor – in my case, for the first time today. It is both comforting and calming, and, thanks to the way that Turner makes the link between physical and emotional surrender as we move, the experience is more poignant than I would have expected.

Then he teaches us a short routine that initially seems impossible but which soon has us all breaking into satisfied grins as we finally get it right. But the most telling exercise of all is when we work in pairs to play 'follow my leader'. The pair touch their fingertips together and one partner closes their eyes. The leader then gently moves their partner around the room, changing speed and direction, taking them as high as they can reach or down to the floor. Then we swap.

At the start of the session one member of the group, a

young woman of about seventeen, is showing all the embodied signs of anxiety: lowered gaze, shoulders hunched around her ears and a tendency to hug herself tightly when listening to Kevin's instructions. But towards the end I glance over to see her leading her partner confidently around the room. She looks like a different person, smiling broadly, her shoulders relaxed: she looks like she could take on the world. It's an amazing transformation, and according to Turner, the confidence they gain in these sessions is translating to everyday life too.

'They've come on in leaps and bounds,' he says. 'As soon as they walk through the door of my studio their posture changes, they're smiling. I've had reports back that it has affected their school work, their jobs, and their lives in general.'

With so many young people – girls, in particular – struggling with both their mental health and their body image, dance can be a powerful tool to help them bypass the insta-bullshit and start to appreciate the body they've got, from the inside. Turner tells me about one young woman in the group who had serious body confidence issues. After a few weeks in the group she finally plucked up the courage to go swimming again. Research suggests that this is not just a one-off. Young women who focus on their external appearance are at greater risk of depression, but dance has been shown time and again to improve both body image and mental well-being.

To be able to 'dance your feelings', though, it helps to know what they are, and here too dance can help. As a species cursed by our (possibly unique) gift for self-knowledge, some believe that, rather than examining our thoughts, we can better access our internal emotional worlds by tuning in to our bodies.

The problem, though, is that a worrying proportion of people are unable to do this. As many as 10 per cent of women and up to 17 per cent of men struggle to identify how their emotions feel in their bodies and to put them into words. Ten per cent is not to be sniffed at – it's roughly the same proportion of people who have dyslexia in the UK. This is a phenomenon called alexithymia, and while it is considered to be a personality trait rather than a clinical disorder, missing out on this important channel of communication can lead to mental ill health.[19] Depression, anxiety, ADHD and eating disorders have all been linked to it, as well as chronic pain disorders such as fibromyalgia, where there is no obvious physical cause for the symptoms.[20] It's not too much of a stretch to imagine that it could affect many people severely enough to play into the epidemic of stress and mental ill health that plagues modern life.

Old body, new tricks

Getting back in touch with the embodied portion of our emotions could turn out to be the balm that our troubled society sorely needs. In fact, dancer and fledgling neuroscientist Rebecca Barnstaple is betting her PhD on it.

She is convinced that tuning in to your emotions through dance is far from a luxury, or even something that we should consider doing as a hobby in case it brightens our day. Instead she believes that it is essential for running our emotional lives effectively.

She points to research suggesting that dance does

something fundamental to our physiology, allowing us to process bodily alarm signals and return to a biologically balanced state, free from stress hormones, and with the chemical hallmarks of well-being flowing through our veins instead. In one such study several weeks of dance movement therapy improved not only emotional health in a sample of teenage girls with mild depression but also significantly reduced stress hormone levels while increasing serotonin levels, a lack of which is linked to depression.[21]

To a certain extent, these changes could apply to doing any kind of exercise. But Barnstaple says that many studies suggest that dance has the edge in improving mood, self-perception and general confidence. And that, she says, is because moving in new ways helps us to practise new ways of responding to situations that may have come up in the past, or which we are concerned will happen in the future. It's like talking therapy, but using body language instead of words.

'It's different when you embody something,' she tells me. 'There's an intimacy and an immediacy to embodying how you're feeling that is quite different from talking about it.'

This kind of dancing doesn't need a strong beat or even any music. The important thing, says Barnstaple, is to focus on the movements you are making. When done with focus, she says, even walking can be considered dance. In the same way that meditation focuses attention on the breath – something that is usually automatic – focusing attention on movements takes us off autopilot and forces us to make some decisions about how we act, physically. The benefit of this, she says, is that once we have some new ways of moving under our belt, it opens up new ways of dealing with thoughts, feelings and emotions.

More than that, though, says Barnstaple, dance provides a safe space to try out new ways of responding to emotions. 'It's about expanding our repertoire,' she says. 'We have an infinite repertoire of movement, and to just open it up even a little bit, it's like it suggests a new possibility. It's expanding your range, quite literally.'

On the other hand, by sitting around all day, only moving our fingers and thumbs, we risk going through life with only a fraction of the physical repertoire that's possible, and lacking the skills needed to be the best versions of ourselves.

More expressive forms of dance can also bring opportunities to experiment with new ways of behaving, safely, and without consequences. Someone who naturally responds to confrontation by shrinking away, for example, can dance a scenario in which they stand up for themselves without fear. This, says Barnstaple, gives people more options in their behavioural repertoire, which they can consider trying in real life. And if cowering during a past traumatic event, such as a street robbery, has left a person feeling vulnerable, re-enacting it can give a much-needed sense of control over what happened to them, she says.

Similar principles are used in dance therapy, Barnstaple tells me. 'A classic dance therapy exercise would be to give someone a three-part idea to try using a movement phrase, something like "I was, I am, I will be", expressed in three movement pieces,' she says. Once a person has put their experience into dance, they have the option to virtually go back and rewrite history, either changing the event itself or their reaction to it. This can offer new ways to experience trauma that are more constructive than going over old ground in talk therapy.

In a way, dance therapy is the exact opposite of mindfulness meditation. In mindfulness, the emphasis is on noticing thoughts and emotions, but not engaging with them or trying to change them. Not only can dance magnify emotions in movement, it also offers a chance to change your reaction to them to make it what you want it to be.

Becoming more tuned in to emotional experience through dance can have other benefits too. Research suggests that dance makes people better able to read their own emotions and the emotions being expressed by other people.[22] Improving personal and social emotional literacy could help people both drastically improve their own mental health and build more positive relationships to help deal with future challenges.

There remains, however, the fundamental fact that many people would rather stick forks in their eyes than express their feelings through dance. Yet even this isn't necessarily a problem, according to the work of Tal Shafir, a neuroscientist and dancer at the University of Haifa in Israel. In her studies Shafir analyses the way that certain movements are linked to basic emotions such as happiness and sadness – one of the basic tenets of embodied cognition. In theory, she believes, as long as you do some of these movements at some point in your day, it doesn't matter whether you dance or not.

Shafir has found, for example, that 'happy' movements tend to involve lightness on the feet, expansive movements where we reach up and out or jump up and down, and repetitive rhythmic movements. In experiments, getting people to make these kinds of movements for just two minutes significantly improved their mood. And dancing the Hava Nagila, a Jewish traditional dance that involves all of these things, was significantly better at relieving depressive symptoms than riding an exercise bike for the same amount of time. Even just stretching in your chair, or deliberately walking lightly during your lunch break, may go a long way to helping you get through an otherwise tough day.

Feel the noise

Failing that, there's always the tried and tested remedy of leaping around to loud music in your kitchen. It may not look pretty, but it certainly makes you feel good in ways that outstrip other ways of getting the heart pumping.[23]

Strangely, at least some of this feel-good factor may come from the enjoyment we get from simply not falling over. According to Neil Todd, a jazz musician who was formerly a neuroscientist at the University of Manchester, it all comes down to the balance organs of the inner ear.

Balance is taken care of by the vestibular system, a system of three tubes filled with fluid, which sloshes around over sensitive hairs when the head is nodding, shaking from side to side or tilting to the left and right. This information is combined with inputs from a pair of 'ear stones', or otoliths (one called the saccule, the other the utricle), which monitor the effects of gravity and tell us whether we are propelling forwards and backwards or moving up and down (see fig. opposite).

Studies of fossilised vestibular systems from early humans have shown that, as we spent more and more time tottering on two legs, the inner ear gradually changed size and shape to become more sensitive to toppling forwards

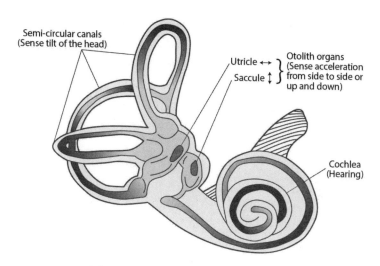

Semi-circular canals
(Sense tilt of the head)

Utricle ↔
Saccule ↕

Otolith organs
(Sense acceleration
from side to side or
up and down)

Cochlea
(Hearing)

The vestibular system of the inner ear

and to the side, with larger loops in two of the three semi-circular canals. This increased sensitivity to falling may have inadvertently fed into our love of dance.[24]

This is because, as Todd points out, the inner ear is wired directly to the limbic system, brain circuitry that is involved in the sensation of pleasure. This, he says, is why we like things such as swings, roller-coasters or freewheeling a bike down a hill, all of which involve hurtling around at top speed. The urge to shout 'whoo-hoooo!' as we fly through the air comes from the tight links between our super-sensitive vestibular system and the brain's pleasure zone. With this in mind, says Todd, it shouldn't be that surprising that we enjoy moving our bodies from side to side and up and down. Nothing feels better than taking your balance organs for a ride, and once you've had that feeling, you just want to do it again and again.

There might be something else going on too. As a general rule, when we go out to dance, the music has to be loud, and this, Todd says, is because above a certain decibel level it hits us right in the ear stones.

The otoliths are an ancient part of the inner ear which once doubled up as organs of hearing as well as balance. Fish and amphibians, for example, still hear via vibrations of their otoliths. In many animals, over evolutionary time, the cochlea took over the job of hearing and the otoliths specialised in gravity detection instead. But Todd believes that our otoliths can still hear, particularly at low frequencies and particularly when those sounds are played at over 90 decibels. This magic number has become known as the 'rock 'n' roll threshold' because below this the music just doesn't seem to get people moving.[25] Sure enough, measurements of music at rock concerts and dance clubs have put the decibel level at between 90 and 130 decibels, with most of the action happening in the low-frequency range.

One of the otoliths, the saccule, which detects up-and-down movement, seems to be particularly sensitive to sound, says Todd, which may explain why, at the very least, we feel the need to bob our heads or tap a foot to the music, and why heavy drum beats have us stomping up and down on the dance floor instead of artistically wafting around.

Groovy syncopated beats are even more fun because they briefly knock us off balance, forcing us to right ourselves. The idea here is that, much as a joke makes us laugh because it violates our expectations and we laugh when we realise it is all a ruse, being wrong about the beat kicks off a short-lived fight-or-flight response, a spike of unease that is swiftly replaced with a wave of relief when we realise that all is well.

Todd says that this too comes down to the vestibular system and the way it helps us keep upright. 'We could think of a syncopation as being a bit like a trigger for a reflex response to a stumble during locomotion,' he says.[26] In other words, if human walking is best described as 'controlled falling', dancing is even more so, and it feels great to save ourselves over and over again.

How to move: dance

- **Stomp to the beat:** Or nod your head or punch the air in time. Following the beat with the body gives us a mood-boosting hit of dopamine that leaves us feeling good and wanting more. Really go for it and you might find yourself catapulted into an altered state of consciousness.
- **Synchronise:** Moving in time with others, whether in a dance class or in the group exercise of your choice, blurs the lines that our brains draw between 'us' and 'them'. This brings us closer together physically and emotionally and makes us more likely to co-operate.
- **Fairy feet:** When you're feeling stressed, take a break and either go for a springy walk or practise jumping and landing silently. Research suggests that moving lightly on your feet is a fast way to lift mood. Bonus happiness is up for grabs if you also put your hands in the air and wave them like you just don't care.
- **Throw yourself off balance:** Do a cartwheel, ride your bike down a bumpy hill or shake your head while you dance. The balance system of the inner ear is linked to the pleasure centres of the brain, and the sensation of almost (but not quite) falling is one reason dance makes us happy.

5

Core Benefits

Stand up straight!
Everyone's mother, ever

In 1945 Joseph Pilates, inventor of the famous core-crunching programme, boldly claimed that 'spinal rolling and unrolling exercises' would 'relax the nerves' and eliminate the '"poisons" generated by nervousness'.[1] Until recently, Peter Strick would have scoffed at such nonsense. A neuroscientist and lifelong worry-wart, Strick is well aware that his tendency to dwell on past mistakes is far from good for his health. But, having spent years tracing neural pathways back and forth between the body and brain, he could see no biological basis for the idea that tightening his natural corset would do anything whatsoever to help.

'My kids would say, maybe you should try Pilates or yoga to help reduce your stress,' he says. 'I'd say, c'mon, give me a break. But then,' he adds with a Harrison Ford half-smile, 'I did this research.'

Strick, a professor of neuroscience at Pittsburgh University, is a serious man, no doubt about it, but his intense focus is softened by his gentle manner and a refreshing honesty about his struggles with stress. In his office in Pittsburgh he introduces me to his emotional-support dog, a giant schnauzer called Milo. Milo clearly takes his work seriously, growling softly when I take a seat between the two of them.

Before we get on to what changed his mind about Pilates, Strick gives me a rundown of his research so far. His life's work – and passion – is to create maps that show how the brain and body are wired together in complex looping neural circuits. It's laborious work and is often considered boring: a kind of neuroscience equivalent of trainspotting. But Strick insists that it is only by tracing the route map of our nervous system and its major interchanges that we can begin to see which parts of the body are talking to each other and to the brain. And only once you know that can you start to work out what is being said and why. He was, in fact, the first to find neural links between the cerebellum, once thought to deal only with movement, and regions of the brain that deal with emotion and cognition.[2] Uncovering hidden links between thinking, feeling and moving is very much his thing.

After several decades of work in this area he was confident that there was no scientific reason why contorting his body into unnatural positions would do anything other than distract him from his troubles for an hour or so. But a surprise finding from a recent set of experiments persuaded him to rethink this position. In 2016, almost by accident, he and his research team discovered a neural pathway that connects the control of the movements of our core muscles to the adrenal glands, the first line of the body's stress response.

The finding could help to settle arguments over a much-maligned area of psychological research into how posture is linked to state of mind. It may also provide a biological basis for why exercise involving the core, such as Pilates, yoga and tai chi, seems to alleviate stress, depression and so-called 'psychosomatic' illnesses – those that have no obvious physical cause and are all too often dismissed as 'all in the head'.

Strick, incidentally, has no time for this particular cop-out. He likes to quote a line from *Harry Potter and the Deathly Hallows*, in which Harry talks to the ghost of his headmaster during a near-death experience: Harry asks, 'Is this real? Or has this been happening inside my head?' and Dumbledore replies, 'Of course it is happening inside your head, Harry, but why on earth should that mean that it is not real?' 'That's the point,' says Strick. 'These circuits are real.'

Professors Strick and Dumbledore are not the only ones making this connection. Neuroscientists who have spent years explaining the mind as a kind of black box that takes inputs, processes them and spits out meaning, are starting to consider that to really understand the mind they need to factor in what's going on below the neck.

There isn't a handy off-the-shelf phrase that encapsulates this brave new world, partly because it's difficult to link mind and body without using words like 'holistic', which, while accurate, have been tarnished by decades of New Age waffle. Chatting to neuroscientist Micah Allen, who studies how the body's internal signals feed into consciousness, it seems I'm not alone in this dilemma. He tells me he describes what he studies as 'brain–body interaction', but admits that it's not perfect because it still sounds like the brain and body are two separate entities. Still, he says, it's an attempt to move away from a 'purely old school, input–output kind of idea, to something more dynamic and embodied'.

In this new view of consciousness as a mind–body

phenomenon there seems to be something special about the core. For a start, it's the area of the body where almost all of our internal organs are found, which means it's the point of origin of many of the interoceptive messages that update the brain about how things are going inside the body. Catherine Tallon-Baudry, a neuroscientist at the École des Neurosciences in Paris, speculates that our organs' central location may actually be the reason why we have a first-person view on the world. Her theory is that the sense that there is an 'I' looking out from the centre of our bodies comes from the body's monitoring of unconscious visceral sensations from the heart and gut. Both generate their own electrical rhythm, independently of the brain, which could act as a constant 'ticking clock' in the centre of the body, providing a reliable reference point for us to hang our sense of self on to.[3]

The trunk, and specifically the core muscles, are also smack bang in the body's centre of gravity, which is why – as any Pilates teacher will tell you – they are so crucial for posture and balance. Even when we're not moving, the muscles of the core are in a state of low-level contraction, which keeps the upper body upright whenever we're not slouching under our weight or leaning against anything. And when we move, the core adapts to keep the mid-section stable so that we can use our limbs to explore and interact with the world without toppling over.

Because this 'bracing' function happens automatically, for a long time thinking was not thought to have any input into the equation. But recent experiments have shown that physical and mental balance have more to do with each other than you might expect.

In experiments where people were asked to think on

their feet – literally, by doing a test a bit like a more scientific version of 'Where's Wally' – researchers found that,
when standing, healthy people engage their core and other
muscles to reduce the sway of their upper body, which allows
them to focus their eyes – and mind – on the task.[4] Similarly,
having to concentrate on staying upright (while walking on
tricky terrain, for example) incurs a slight cost to cognition.

This only really becomes a problem when either cognition or posture is compromised to the point where it becomes
impossible to either stay upright while you're thinking or
vice versa. Falls are second only to road accidents as the
most common cause of accidental death worldwide, and,
while they disproportionately affect people over sixty, the
worrying thing is that we are starting to lose our balance
surprisingly early.[5] In a study of over a thousand people,
women's ability to balance peaked in their thirties before
starting to gradually decline. Men start to lose it even
earlier, between twenty and twenty-nine, although they
seem to have better balance to start with (perhaps because
they tend to have more muscle mass).[6] Cognitive skills too
start to slide as we leave the heady days of young adulthood
behind. Despite what brain-training companies tell you, the
best evidence on how to protect cognition into old age is not
to think more but to stay as physically active as possible. At
least some of this effect may come from the fact that any
kind of movement strengthens the balance function of the
core.

It may also explain why some studies have found that
posture-related exercise, such as tai chi, both improves
cognition and reduces fall risk in the elderly, perhaps by
cutting the cost of thinking on our ability to stay upright

and balanced.[7] All things considered, it seems sensible to try to keep the core strong from middle age onwards at the very least.

Finding a balance

As for the link between the core and emotional control, Strick's main area of study, there is a growing body of evidence that emotional balance also feeds into physical stability. Studies with the elderly, for example, have shown that fear of falling is itself the number one risk factor for falling, partly because it changes posture, making it more stooped and less balanced. Not only that, but psychological experiments have been showing for years that posture matters when it comes to state of mind. Mental illnesses, including anxiety, depression and schizophrenia, have all been linked to changes in posture that increase the risk of falling.[8]

Nailing down the physical connections between activation of the core muscles and emotional responses is important because, despite a huge number of psychology studies pointing to the link, nobody had found a convincing mechanism to explain why an upright posture is linked to feeling positive, powerful and in control, while, on the other hand, being hunched makes you feel as defeated as you look. Without this killer piece of the puzzle it was all too easy to dismiss the emotional benefits of standing up straight as the effects of expectation (you've always been told do it so it must be 'good') or, at worst, flimsy pseudoscience spawned from bad studies.

Social psychologist Amy Cuddy found this to her cost. Back in 2012 she was a researcher at Harvard Business

School when she made a big splash with a TED talk based on her research into what she called 'power posing'. In her experiments, conducted with Dana Carney and Andy Yap, both then at Columbia University in New York, she asked one group of people to hold 'expansive' postures, in which they took up as much space as possible: standing with their legs apart and arms in the air, or leaning back in a chair with their feet up on the desk, for two minutes. A second group sat hunched in their chairs or with their arms folded. Afterwards, the power-posing group reported feeling more powerful and performed better under stress than the ones who had sat slumped in their chair for two minutes. In the initial study the team explained this by showing that power posing decreased levels of the stress hormone, cortisol, in the blood, while boosting testosterone levels.[9]

The idea caught on, and the TED talk became the second most-watched of all time. The media ran with the story in a big way. Cuddy became a bestselling author and a sought-after motivational speaker around the world. The problem was that when other psychologists repeated these experiments they didn't find the same hormonal changes, and the whole thing started to unravel. The backlash was brutal. Cuddy was battered by her peers for peddling flawed science and for seeking fame off the back of one study. Even Dana Carney, who was the lead author on the original study, disowned the research and said that, in her opinion, the results were 'not real'.

These days Carney refuses even to discuss the research in the media.[10] Even so, as is often the way with wild swings of opinion, among many psychologists the needle has begun to drift quietly back towards the middle. The current status quo seems to be that posture matters, and though we don't exactly know why, from what we can tell it's probably not the hormones.

In any case, a recent review of the field of research has seemingly vindicated the part of the research that matters to anyone who wants to use it: that taking an expansive posture does appear to make people feel more powerful.[11] Perhaps unsurprisingly, Cuddy herself has stepped away from the whole area and now focuses her attention on the effects of adult-to-adult bullying.

While the power-posing poster girl has moved on to other things, other psychologists are continuing to report that standing – or sitting – up straight can make the difference between feeling in control and wanting to hide under the duvet. Slouching – what Cuddy would call a 'contractive' posture – has meanwhile been linked to feelings of defeat, social withdrawal and fatigue.

The same kinds of posture – open chest and puffed up versus slumped and tucked in – are seen in other social animals too. This suggests that it is probably an innate rather than a learned behaviour. To this day, slouching works as a social signal – broadcasting 'I give up' to rivals and 'help me' to supporters in our hour of need. The great thing about being human, though, is that we have the capacity for metacognition: to reflect on our actions, thoughts and feelings and change them for the better. By noticing your posture and deliberately adjusting it, it's possible to hack into this automatic system, change your posture and change the message that's being sent to the 'how I feel now' chatroom of the brain.

To this end, Elizabeth Broadbent, a health psychologist at

the University of Auckland in New Zealand, has been investigating how changing posture may change the biological way that we deal with stress. Previous work had shown that people sitting in a slouched posture find it easier to remember negative words from a list, while sitting up straight favoured recall of positive words. With this in mind, she put volunteers through a standardised version of everyone's worst nightmare: having to write and then present a speech at short notice to a group of judgemental-looking strangers. This reliably increases heart rate, blood pressure and sweaty palms and, if a person happens to be in a bad mood already, it makes them feel considerably worse.

But, Broadbent's studies suggest, sitting or standing up straight provides a buffer against this kind of stress. It promoted a more positive mental attitude, borne out by higher mood ratings, lower fatigue and less anxiety. Slumping led to the opposite – people reported feeling miserable, defeated and lacking in energy. What's more, when she and her colleagues analysed the content of the volunteers' speeches, they found that people in the 'good posture' group also spoke in the first person less often, suggesting that they were less focused on themselves. This is striking, because a tendency to focus inwards is a feature of depression: it is linked to a tendency to beat yourself up and dwell on past mistakes.

In a separate study, also by Broadbent's group, people were given the same stressful test while walking upright on a treadmill or while looking at their feet. This time, their physiological state was also measured. Under normal circumstances, giving a speech at short notice reliably increases heart rate and blood pressure and makes people sweat more. And yet, the study showed, doing so while walking with a straight back and head up significantly reduced blood pressure and sweating, compared with a slumped group. It also made the test subjects feel more alert and less tired. This particular experiment couldn't say whether it was because standing up straight alone reduces blood pressure by default or because of the impact of standing up on the stress response, or both. Nevertheless, upright people also had much lower sweat levels in the recovery period after the speech, suggesting that they bounced back faster than the stressed-out slouchers. And, whatever the precise mechanism, the finding that standing up straight buffers stress is something that we can all put to use very easily.

Broadbent speculates that there is almost certainly more than one thing going on here. While she hasn't yet tested this in a study, a major feature of a slouched posture – and of depression – is that the gaze tends to be directed to the floor. This clearly affects what you see and what you have the opportunity to respond to and, she speculates, may direct your focus inwards. Simply looking around at the world automatically demands more interaction. And then there are the probable physical effects of slumping on the heart, lungs and pipework of the body, which may affect blood pressure and the amount of oxygen being pumped around and have knock-on effects on energy levels.

While Broadbent stops short of giving a mechanism for her findings ('I think it always helps to be a bit modest in your claims until you've got enough research to back you up,' she observes), the take-home message here is clear: the bad times will feel more manageable if you stand up straight and look the world in the face.

There are more organised ways to get the same thing,

of course. Upright and expansive postures are a key feature of yoga and tai chi, both of which are key areas of focus for Peter Wayne, who leads the Osher Center for Integrative Medicine at Harvard Medical School. Wayne is both a tai chi instructor and a researcher, first in evolutionary biology and more recently in holistic medicine. In his early career he was taught by the legendary biologist E. O. Wilson, who in a lecture on the evolution of body language showed a series of images of people from around the world doing expansive 'victory' postures. 'It just made me cry because I was already teaching tai chi,' says Wayne. 'I started thinking maybe that's why there's all these shapes in tai chi. Maybe these shapes that we do in yoga encode a certain quality.'

In a recent analysis of embodied cognition and movement Wayne concluded that they do – specifically, by lifting mood and bringing a sense of calm focus. He quotes the Zen master Shunryu Suzuki: 'These forms are not the means of obtaining a right state of mind. To take this posture is itself to have the right state of mind.' [12]

*

Back in Peter Strick's lab, where Milo has stopped growling but is still keeping an eye on me through his big, black, bushy eyebrows, Strick thinks that his team is on to something important that could not only go a long way towards explaining why posture matters for our state of mind but may also provide tools for us all to deal with the stresses of modern life a little better.

He came to study the stress system almost by accident, after spending many years mapping the motor cortex. This is a strip of brain tissue that sits like an Alice band over the top of the head and is in charge of sending messages to the muscles when it's time to move. In 2012 a gastroenterologist called David Levinthal joined the research team with an interest in using Strick's neural tracing methods to find out how stress might affect the health of the gut. At the time, movement wasn't in the mix. All Levinthal wanted to know was why so many gut problems are exacerbated by stress, and he planned to find out by tracing neurons involved in the stress response back to the brain, to see if they end up in regions that then go on to update the stomach.

As a lifelong worrier, Strick was intrigued by this line of inquiry. 'As a kid, I had stomach aches, and my parents took me to a doctor,' he recalls. 'The doctor said, "There's nothing wrong with him, he's got some psychosomatic problem, it's all in his head".' Collaborating on Levinthal's project offered Strick the perfect opportunity to find out whether that dismissive doctor was right, by targeting the adrenal glands, part of the stress system, and tracing the neural wiring back to the brain – which, as Strick points out, is indeed all in the head.

The adrenal glands sit on top of the kidneys and are responsible for pumping out adrenaline, which drives the fight-or-flight response. Most of the fight-or-flight action happens in the central part of the gland, the adrenal medulla, which is made up of modified nerve cells that not only release adrenaline into the blood but also have a direct line – made of superfast neural cable – to the spinal cord and on to the brain.

Tracing neural pathways is a fiddly process, and there are good reasons why it isn't done more often. It involves injecting a virus into an organ of interest – ideally a virus that only infects neurons – and waiting for it to spread through the nervous system on its way back to the brain. Later, the brain samples can be tagged with a marker that shows where the virus ended up.

After years of trying with various viruses that infect neurons, including polio and several strains of the cold sore virus, Strick and his colleagues found that the best virus for the job is rabies – a disease that uses the body's nerve pathways as a fast track from the site of entry into the body and on to the spinal cord, from where it can head on up to the brain. It takes days – sometimes weeks – for the infection to move along a chain of nerves, crossing the connections between individual cells as it jumps from one neuron to the next. By using a particular strain of rabies that only infects neurons and leaves the surrounding tissue untouched it is possible to get a nice clean view of where a particular pathway leads. Clearly, using rabies – for which there is no cure – rules out human experiments.

There's no way of sugar-coating this: several monkeys died in the making of these discoveries. I am an animal lover, as is Strick, and we discuss the ethics of all of this many times during my stay in Pittsburgh. To summarise where we got to, there are no easy answers. The same arguments in support of studying monkeys – that their brains are very similar to ours – is the same argument that can be made for not experimenting on them. Rat studies would be less controversial, but, Strick points out, they would also be pointless because rodents lack so many of the more specialised areas of cortex that we have. It's like the old story of the drunk man who is looking for his lost keys under a street light, he says. When a passer-by stops to help and asks where he dropped them, the drunk man says 'over there in the park'. 'So why are you looking for them here?' asks the passer-by, to which the man replies 'Because this is where I can see.'

'We can do rodent research and not learn anything,' says Strick. 'And that's an option, but it's looking for answers underneath the lamplight. If you want to learn something about these systems, responsible experiments in non-human primates are about the only option.'

Ultimately, it's a value judgement that can only be made after weighing up whether you think the end justifies the means. Strick is at pains to point out that the monkeys don't suffer any symptoms of rabies. While the virus is passing through the nervous system it gives no clues to its presence for up to several weeks. He also points out that we have almost no cures for neurological diseases, and that's partly because we don't fully understand how the nervous system is wired up. 'I have understanding for people who feel that we shouldn't be doing this,' he says. 'At the same time, I think it's essential to improve the human condition.'

Stressed to the core

Because so many of us spend our lives in a state of mild anxiety, it's easy to underestimate the effects that stress has on the human condition. Chronic stress is linked to an increased risk of almost every life-threatening disease, from heart disease to cancer, Alzheimer's and depression. Then there are the social and economic costs of burn-out, addiction and crime, all of which are in one way or another linked to people becoming overwhelmed by life's challenges.

Finding the physiological pathways that control it could make proper evidence-based stress management not only more accessible but also vastly more effective.

With that in mind, the finding that the adrenal medulla is linked to movement-related areas of the brain is as important as it is surprising. It offers an alternative route to stress control that has nothing to do with changing the way you think or trying to rewire deeply entrenched emotional habits. It also suggests that, rather than paying lip-service to exercise as an add-on to other forms of therapy, we should perhaps start thinking of it as one of the key pillars of good mental health, with movement given the same importance as mind-based interventions such as mindfulness meditation and cognitive behavioural therapy.

As for the kind of movement we should be doing, when Strick looked at the motor cortex of the monkey's brains, the links to and from the adrenal medulla were overwhelmingly in the part of the brain that drives the core. 'There's something about movement, no question about that,' he says. 'And, activating your core has an impact on the adrenal medulla more than other places.'

The stress-control loop doesn't begin and end with the core, however. Clearly we have other ways of calming down in challenging situations than physically moving away. A significant proportion of connections to the adrenal medulla come from cognitive (thinking) regions of the brain, particularly those that help us make sense of conflicting information. These regions may well come into play when we think ourselves out of trouble, or reason our way out of anxiety. Similarly, emotion-based regions of the prefrontal cortex that are known to be activated during mindful meditation are also connected to the adrenal medulla, which may explain how mindfulness can mitigate stress in the moment.

Interestingly, a brain region that processes sensory information coming from the back also talks to the stress system, which may explain why patting or rubbing a crying baby's back calms them and sends them off to sleep, and why a back massage feels so incredibly relaxing.

A few links to the adrenal glands also appeared in parts of the motor cortex that control the muscles around the face and eyes – the ones that are activated by a genuine smile, the type that gives you twinkling eyes and crow's feet. In experiments, people were tricked into contracting these muscles while doing a really annoying puzzle – the type that sounds easy until you actually try it. Their task was to trace the outline of a star on a piece of paper, using their non-dominant hand, as many times as possible within two minutes, with a prize of chocolate for the winner. The catch was that the paper was hidden inside a box, and the only way to see their hand moving was to look at it in a mirror. To make matters worse, volunteers had been told that most people manage to trace eight stars in two minutes, with fewer than twenty-five errors. This wasn't true: the average was just two, with over twenty-five errors. The study found that, in spite of the rage-inducing nature of the task, people who smiled all the way to their eyes throughout suffered less stress and recovered more quickly (measured via their heart rate) than those who only showed their teeth.[13]

This suggests that mental and emotional stress may be treatable not just by thinking or talking about your problems, or even by taking time out to relax, but also by moving in a way that activates core muscles and, ideally, by doing something that makes you properly smile. Break into a full belly laugh and you hit two pillars of stress control in one go: a recent study found that laughing really hard provides a better core workout than crunches.[14] Or, if your friends aren't that funny, it might be worth considering laughter yoga, which uses breathing exercises and other movements to work the same muscles that are used in a real belly laugh. It sounds mortifying, but according to the few studies that have been done on it, it actually works. Forcing yourself to laugh hard changes your physiology in the same way as real hilarity, making you feel happier in the process.[15] Laughter yoga has also been shown to help reduce anxiety and stress in a way that makes it a potentially useful add-on to depression treatment.[16]

FYI

As for the all-important mechanism by which core activation works to turn things around, it may be the biological equivalent of an FYI to the adrenal medulla, that the body is on the move and that some basic physiological housekeeping needs to be taken care of.

This comes neatly back to the reason that we have a brain in the first place: to allow us to move our bodies in a way that allows us to take appropriate action. In stressful situations this could mean a full-on fight-or-flight reaction, stimulating enough adrenaline to fuel running away or fighting for your life. Or, in less dramatic circumstances, it could send a lower-level signal to fire up the muscles because the body is on the move.

Move!

Any kind of movement involves input from the body's sympathetic nervous system – the branch that tweaks the width of the blood vessels to adjust blood flow, adjusts heart rate and does all the behind-the-scenes action that allows us to move. The connections run in both directions, meaning that the strength of the response is constantly being adjusted to meet the needs of the body and the action it needs to take to survive.

An important foundation of any movement is stabilising the trunk, for the simple reason that it provides the base from which the limbs can move. 'If you reach and you're standing, if you didn't contract your postural muscles, you wouldn't stay standing,' says Strick. And, to add a particularly nice mental image to the mix, he adds that almost any movement also involves contracting the muscles of the pelvic floor (also part of the core) because, as he puts it, 'you don't want your intestines going out of your rear'.

The only snag is that, with the kind of neural tracing that Strick does, it isn't possible to tell whether the signal to the adrenal glands is saying 'rev up', 'ease off' or a mixture of the two. But, says Strick, we already know that posture affects the way we feel, thanks to all of those psychology studies. 'We have lots of clues about what that action might be,' he says. 'When you see someone who's depressed, their posture sucks. And there's something about upright posture that has some positive influence,' he says.

In an annoying, journalist kind of way I push him for more. Maybe it's a 'calm down' signal or a switching off of the revving-up signal, I suggest. 'I'd like to say that your insights about this are probably as good as mine,' he replies. 'And that's OK. But what I can tell you is that this region has more impact on the adrenal medulla, and we have lots of evidence that there is something about the core as a stress-reducing feature. And I think that's good enough.' For now, at least.

Sit up straight and smile

The good news is that, although huge numbers of people are neglecting movement as a route to stress reduction by sitting around all day, the fact that core muscles are involved in everything means that, however you choose to move, it should help alleviate your stress levels. Adding in some core-blasting exercise in times of particular stress might help send a message back up those wires that the body has moved out of danger and the whole stress cascade can stand down. And keeping the core muscles in good condition should help stave off a middle-aged decline in balance.

Tweaking everyday habits could bring further benefits. Sitting may be the new smoking, but ways of sitting that engage the core – whether that's sitting up straight, kneeling or squatting on a medicine ball – are likely to be vastly better than slouching on the sofa with a laptop on your knees (guilty). There's also walking and *walking*. Keeping the head up, spine up and smiling to all and sundry might get you a few strange looks in London or New York, but it might also go a long way to keeping the stress of city life in check.

People with stress-related illnesses might get even more benefits. David Levinthal, Strick's gastroenterologist colleague in Pittsburgh who got the team thinking along these lines in the first place, tells me that it increasingly looks as though what used to be labelled 'psychosomatic' gut disorders are actually a dysfunction of the mind–body interaction, and that working on the core muscles may help.

One interesting finding in this regard is that in healthy people the core muscles automatically contract to offset changes in pressure inside the abdomen as the gut goes about its business of moving food and gas through the system. In some people with irritable bowel syndrome this reflex doesn't work properly, Levinthal tells me, which may help to explain why IBS so often causes bloating. Clinical trials of yoga for IBS have shown promising results, relieving both gut symptoms and anxiety, which backs up the idea that core exercise might be of value to tackle both stress and gastrointestinal issues, perhaps in part by developing a stronger core. This could help in several ways: not only help reduce stress by activating the body–mind neural pathways, but also by strengthening the muscles that are in charge of pulling in the belly.[17]

'Clinically, I think that yoga, tai chi, Pilates are part of a larger intervention for stress reduction,' says Levinthal. 'I think there's an emerging view that core muscle exercises can play a role.'

Connected

All things considered, it is starting to seem as if the core provides common ground that links many different ways of moving, whether dancing, walking or learning to backflip, to emotional control. One core muscle in particular keeps coming up: the psoas, which connects the spine to the top of the thigh bone, and is intimately intertwined with the diaphragm. It's also the muscle that pulls the legs up and forward when we walk or run.

Because of its position in the body, connecting breath with movement, the word on the street in yoga, Pilates and a dance-based combination of the two called gyrotonics is that the psoas is *the* muscle connecting the stress response to the physical act of running away and the extra-deep breaths that are needed to keep it up. And, since the psoas is shortened by too much sitting, the theory goes, it's hardly surprising that we are all so stressed: we are constantly in a state of half-formed fight-or-flight.

So far this is based on a lot of speculation and very little research. But given what Strick has found about the link between stress responses and the core, it is certainly intriguing. It could be that stretching and oiling the psoas by moving more, and strengthening it along with the rest of the core will help to build a healthier, more adaptable stress response.

What's needed now are specific studies to look at core-strengthening as a specific form of stress control in comparison to other interventions. So far, there have been a few that show improvements in mental health with Pilates,[18] but as yet this can't be pinned only on the core-stability element as opposed to the breathing, me-time and the calming influence of a teacher who cares.

David Levinthal is the first to admit that more work is needed. But, he adds, 'it seems too much of a coincidence that the sites in the brain that control the abdominal muscles are smack dab in the middle of this representation of the organs.'

While we wait for the final piece of the puzzle, though,

there is, to my mind, already enough evidence to suggest that it's worth putting in the effort to make these key muscles as strong as they can be. At the very least it will improve posture, which we know has almost instant effects on mood and cognition. And there's no harm in strengthening the body area that provides a base of support for other aspects of life-enhancing movement. Whether you do it via yoga, dance, walking or while lying on your back at the gym, it's starting to look very much like it might be crunch time.

How to move: the core

- **Work the core:** Running, Pilates, yoga, swimming, it doesn't matter how you move. Any way of moving that activates the core muscles sends a message connected to the adrenal glands via the brain to help regulate stress. We don't yet know exactly how, but engaging the core seems to tell the body to calm down.
- **Have a laugh:** A deep belly laugh works the core muscles more effectively than sit-ups, and its stress-busting properties extend beyond the core-to-brain pathways. Smile to your eyes while you do it for a second hit of stress relief.
- **Sit up, stand up:** It's harder to think positive when you're slouching, and sitting or standing up straight brings more positive thoughts. Keep your head up and gaze forward for even more benefits.

6

Stretch

I bend so I don't break.
Unknown

In a laboratory at Harvard University a rat is doing a downward dog. Its little red eyes are half-closed, and it looks for all the world as if it's enjoying it as much as I do.

A good stretch has got to be one of life's greatest pleasures. It's also about the fastest way to use movement to change the way you feel, particularly after several hours spent scrunched up in a chair.

It's becoming clear that there's more to stretching than just a way to loosen up tense muscles. In fact, research that began with those happy Harvard rats is revealing that it can act as a full body reset with far-reaching effects on mind and body – and may even affect the biological basis of health and well-being.

Strictly speaking, the stretching that happens after a long spell of stillness is a little different from what goes on in the average yoga class. The kind of stretch where your arms make a start on the Y of YMCA, while your shoulder blades pull back and you yawn so wide it feels like your jaw might snap off is called 'pandiculation', and it's not entirely under conscious control. It's so ubiquitous in mammals and some birds that it is believed it may have evolved as a reflex that wakes muscles up after rest – firing up the sensory nerves that run to the brain, reminding it that the muscles are there so that they are primed and ready for movement.

Humans, being humans, have taken this happy act and turned it into something far less fun, something that we feel 'should' be doing before and after we exercise. The answer to whether we should is still quite unclear – scientists disagree over whether stretching makes any difference to a person's chance of injury during exercise, and indeed whether it helps or hinders performance. What's more intriguing in the context of body–mind interactions, though, is that new evidence suggests that stretching changes the physical and chemical properties of the body's tissues right down to the cellular level. This in turn may have ripple effects through-out the whole body, via the immune system, with important implications for both mental and physical health and the all-important links between the two.

The surprising bit about this new role for stretching is that it isn't all about what happens in the muscles. We know that stretching muscles can release them from the state of partial contraction that they get into when we sit around for too long. Tension from sitting tends to disproportionately affect the neck, shoulders and hips, partly because concen-trating makes these muscles work hard to keep the head still so that we can focus, physically and mentally. We also tend to tip our pelvis forward when we sit in a chair, putting strain on the lower back and shortening the muscles at the front of the lower body, including the hip flexors and psoas. Remembering to stretch after a period of sitting can relieve this tightness and, if we do it often enough, may stop stiffness from setting in for the long haul.

The immune link comes from another kind of tissue entirely: fascia – a type of connective tissue that basically holds our bodies together. If you've ever wondered how our organs stay in the right part of the body rather than rattling around knocking into each other, the answer is fascia. It's everywhere. Sheaths of the stuff separate organs into compartments, surround every muscle fibre and artery, and wrap around individual muscles to keep them separate while allowing them to slide over each other as we move. In fact, according to Helene Langevin, who did the rat yoga experiments at Harvard and now researches fascia at the US National Institutes of Health in Bethesda, Maryland, if you were to remove every organ, bone and nerve cell in the body, it would be much the same shape, as long as you left the fascia intact.

Technically, connective tissue is a catch-all term that covers all kinds of things, from bone to blood to fat, cartilage, tendons and skin. All of these tissues have more or less the same basic structure: cells and various proteins sitting in some kind of a matrix. The differences between different types of connective tissues come down to the relative amounts of each ingredient, as well as what else is mixed into the matrix: calcium, in the case of bone and teeth, for example, which makes them hard. Fascia is built from strong collagen and springy elastin fibres woven into strong yet pliable sheets. It varies from multiple thin layers of slimy cling film that lie just beneath the skin to thicker, more fibrous sheets that surround the muscles and line our various body cavities.

Despite the fact that it's everywhere, or perhaps because of that, this kind of material hasn't been as well studied as some of the body's other tissues. From the earliest days of science anatomists would scrape it off and chuck it into the bin so that they could get a look at the interesting stuff beneath. It's not as if they didn't notice it was there – far from it – it's just that sticky, gloopy and sometimes rubbery white stuff is not helpful if you want to see the anatomical wood from the trees. And, frankly, they didn't see the point in spending too much time investigating the properties of what looked like nature's version of cling film. This type of 'loose' connective tissue exists to connect different body parts and wrap them neatly in bags, end of story.

Alternative therapists, on the other hand, have a long history of being fascinated by it. Ida Rolf, who in the 1940s invented a kind of deep tissue manipulation that she called Structural Integration and everyone else calls 'Rolfing', considered fascia to be a vital ingredient in aligning the body's (supposed) 'energy field' against the Earth's (very real) gravitational field. Chiropractors and osteopaths also believe that the fascia can be 'released' through massage and manipulation, making movement more fluid and curing all manner of ills. In recent years 'fascia' has become the latest buzzword in alternative medicine. Bodyworkers, yoga teachers and wellness gurus alike have taken the theme and run with it, weaving the proven and the unprovable seamlessly together until it's difficult to know what to believe.

In the past decade or so scientists have begun to take an interest too. From my standpoint as a professional sceptic, and having tasked myself with extracting truth from fantasy, it's awkward that a large proportion of the scientists who are working in this field either have side practices as Rolfers, chiropractors, yogis or acupuncturists or are at least open to the idea that alternative medicine works. Which, to me, makes it difficult to believe that they are truly objective about what they are studying.

With this in mind, I was a little hesitant about contacting Helene Langevin to find out about her work on the subject. Not long ago she published a research paper plotting the acupuncture meridians (invisible lines of mysterious energy) along some of the body's major fascia junctions; that raised a few eyebrows in scientific circles.[1] On the other hand, she's a very senior scientist, and she was a professor at Harvard University before she became a director at the US National Institutes of Health. I didn't want to dismiss her research without looking into it properly. But the obvious question for me was: does she *really* believe that there are invisible energy lines flowing around the body?

'These things are hypotheses, they are not facts,' she tells me, matter-of-factly. 'It's very important to distinguish between them. It's important to be respectful of the traditions but at the same time to recognise that these are not scientific terms.'

Langevin's interest in acupuncture began in the mid-1980s, when, as a practising medic, she became frustrated by the limited treatments that she could offer patients who came to her with chronic pain. She decided to study acupuncture, so that at least she'd be able to answer patients' questions about it with some degree of knowledge. It was during a practical acupuncture class that she felt something that piqued her interest and ultimately led her to study the science of stretching.

'I was learning how to manipulate the acupuncture needles and my teachers were saying that you have to twirl them round,' she tells me. 'And I could feel something going on when the needles were being manipulated.' What she felt was a slight tugging sensation, as if the needle had grabbed onto something under the skin. For the person being punctured this feels like a dull ache a few centimetres around the needle. In acupuncture circles this is known as 'de qi' or 'obtaining qi'.

A decade or so later Langevin moved from medicine into research and finally had the chance to study what actually happens at the cellular level when the needle breaks through the skin and obtains qi. Looking at samples of rat tissue under the microscope, it was actually very clear: when the acupuncture needle hits the fascia layer that lies directly beneath the skin, it picks up strands of the collagen fibres that give it its strength. Then, when the needle is twisted, these fibres twirl around the needle like spaghetti on a fork, pulling the surrounding tissue a little bit tighter and giving it a very localised stretch in the process.

The really interesting bit, though, was what happened to the sauce that the collagen spaghetti is sitting in. Some types of loose fascia have more densely packed nets of collagen than others, but they all sit in a gloopy layer of slime that allows different layers of the sheets to glide over each other. This slime is secreted by cells called fibroblasts, which also make and maintain the fibres. Langevin's team found that when collagen fibres are twirled around the needle, fibroblasts come along for the ride, changing shape in the process.

The changes, Langevin tells me, don't come as a result of

being passively stretched until they lie flat. Weird as it may sound, cells are capable of moving under their own steam, often in response to changes in mechanical forces imposed on them (for example, when we move), and sometimes as a result of what is going on in and directly around the cells. At the cellular level, movement happens via a cell's internal scaffolding system, called the cytoskeleton, a kind of expanding and contracting road network that ferries molecules around the cell and also gives them their shape and structure. When the cytoskeleton expands and contracts, it can change the cell's shape and size, and this in turn can trigger the release of various signalling molecules in and around the cell. The emerging field of mechanobiology is beginning to investigate these effects and the intriguing possibility that these tiny adjustments add up to biologically important changes in the chemical conversations that cells are having with their neighbours near and far.[2]

Langevin found that, during acupuncture, the cytoskeleton in the fibroblasts rearranges itself in such a way that the cells pull themselves to become flatter and several times longer.[3] As part of this process the cells release a signalling molecule, ATP, into the matrix. Anyone who did biology at school will remember ATP as the molecule that acts as the currency of energy release inside the cell. When it is outside cells, though, it has a different job, including managing the levels of inflammation in the tissues.[4] It also seems to have a side-line in making the connective tissue less stiff and more pliable. 'We saw the tissues actually relaxing at the same time that the cells are releasing ATP,' Langevin told me. Overall, it seems that there's something about pulling the fibres that make up the fascia that wakes up the immune system while making the tissues feel softer and more elastic and possibly actually changing the fundamental character of the tissue itself.

At this point I should probably point out that this isn't evidence that acupuncture works as a cure for any ailment. The only thing it shows for sure is that twisting tissue around needles physically stretches it, and that this in turn changes its structure. This made Langevin wonder: if acupuncture is really just a very localised stretch, do you actually need to put a needle into this tissue to make this happen? Can't you just … stretch? So she tried stretching pieces of rat tissue and the short answer was that, yes, exactly the same thing happens at the cellular level if you stretch it any other way. So she put acupuncture to one side and moved on to the broader – and less controversial – question of how stretching affects connective tissue biology.

The obvious question at this point is, so what if stretching changes the structure of fascia spaghetti or the flavour of its sauce? What do potential changes at the cellular level have to do with our mental state? The answer to that comes back to the chemical changes that happen when the tissue is stretched, and what it tells the rest of the body and brain. It all comes down to inflammation.

Over the past twenty years or so it has become clear that inflammation is the ultimate mind–body phenomenon. It's the part of the immune response that acts as the body's first line of defence against illness or injury – the reason a twisted ankle gets hot and swells, and why your nose gets blocked when you have a cold. The exact response depends on the nature of the infection or injury, but the basic job of inflammation is to flood the area with white blood cells, which gobble up any invading pathogens and repair tissue damage. Then, when the threat has passed, other immune cells release other substances that turn off the inflammation response and return the tissues to normal.

That's the idea, anyway. But, as well as responding to actual emergencies, inflammation also ramps up in response to perceived threat – or, to give it its everyday name, stress. In an evolutionary context this is actually the whole point of stress: to raise the alarm that something is wrong and that it would be a good idea to prepare for a fight. The immune system takes the impending threat seriously and ramps up its activity so that it is ready to deal with any battle injuries.

As with so many aspects of the stress response, what worked for our ancestors isn't helpful today, and our bodies haven't quite got the memo. As a result, everyday or repeated stress can leave us in a state of chronic, low-level inflammation because the 'all-clear' message either doesn't get sent out or is swiftly followed by another reason to sound the alarm.

This matters partly because of the risk to physical health: chronic inflammation is implicated in heart disease, cancer and pretty much every other life-threatening illness you can think of, not to mention chronic pain and Alzheimer's disease. It is also increasingly implicated as the missing link between stress and mental ill health.[5] Everyone has experienced the mental aspect of inflammation at one time or another, usually when fighting a cold or flu. That feeling of lethargy, misery and wanting to be left alone to hide under the duvet is part of a slew of responses known as 'sickness behaviours', which probably evolved to prompt animals to seek out rest and isolation until the injury heals or the infection runs its course. It's a well-documented side-effect of inflammation and, in terms of how it makes you feel, is almost indistinguishable from depression.

This isn't too much of a problem if it goes away after a few days. The trouble is that in modern life it often doesn't. Long-term stress, like being a full-time carer, or repeated daily stresses such as commuting to a job you hate, can leave the body in a constant state of low-level inflammation that never resolves. In fact, contemporary life is positively brimming with features that contribute to inflammation. Loneliness and social rejection have both been found to raise the levels of inflammatory markers in the blood, as has a sedentary lifestyle.[6] Obesity makes matters worse too, not least because inflammatory cytokines – biological messengers that start the inflammation response and keep it going – are stored in body fat. The more fat there is, the bigger and faster and longer-lasting the inflammatory response is likely to be. And if all that wasn't bad enough, inflammation also ramps up with age, playing a role in age-related diseases from heart disease to dementia and cancer. It also accelerates ageing itself.[7]

Given that stress, obesity and ageing are major features of modern life, it would be fantastic news if something as simple and pleasurable as stretching could help.

And there is indeed some evidence that it can. Some studies have found that people who do yoga and tai chi regularly have lower levels of inflammatory markers in their blood, although it's difficult to pull out the effects of stretching from the overall benefits of exercise, breathing and general relaxation.[8] Langevin's team have been trying to tackle this by looking more closely at how stretching alone affects inflammation, both in animal studies and with human volunteers.

The results so far are intriguing. In a study published in 2017 Langevin and her team injected the rats' back muscles with a substance called carrageenan – a carbohydrate found in seaweed that is a common additive in processed foods, and which causes local inflammation if injected under the skin.[9] Forty-eight hours after their injection, half of the rats were encouraged into a downward dog position by lifting them by the tail and allowing them to grasp onto a tiny rail. They would pull against the rail, get comfortable and stretch their backs for all they were worth, looking relaxed and happy while they were at it. The other half were handled for the same amount of time but not given the opportunity to stretch.

The results showed that the area of inflammation was significantly reduced in the rats that had stretched, and the tissue had fewer white blood cells (a sign of immune activity) than the non-stretched rats. Even more importantly, the experiments suggested that stretching the fascia starts the sequence of events that actively turns inflammation off, allowing the tissues to return to normal.

Make it stop

This is important, because the main problem with inflammation isn't that it is constantly being turned on. It's that it doesn't get turned off. It used to be thought that inflammation just petered out as part of the cellular cleaning-up process, but we now know that it is, in fact, an active process, meaning that the body needs to send out a chemical signal to douse the flames.

In the early 2000s Charles Serhan, an immunologist at Harvard Medical School, identified three families of molecules: resolvins, maresins and protectins, all of which the body makes from Omega-3 fats in the diet and which turn off inflammation.[10] Working with Helene Langevin's team, he found that the yogic rats had higher concentrations of resolvins in their tissues than the ones that hadn't been stretched. Stretching the site of injury seemed to tell the tissues that the worst had passed.

It remains to be seen how significant this effect is for overall health. The Harvard team has begun a study in human volunteers, in which they are measuring how stretching affects the levels of inflammation markers and white blood cells. An intriguing possibility is that stretching one part of the body may have benefits throughout the system – if resolvins enter the blood, they might clear up inflammation that has nothing to do with stiff tissues but instead is the result of infection, chronic disease or just the miserable process of getting old. If this is the case, then a regular stretch could act as a reset button that stops a bad day from turning into a runaway stress response that results in chronic illness.

One of the questions the researchers hope to answer concerns how long you need to hold a stretch before the cellular changes take place. In the rat studies they held each stretch for ten minutes, but it's possible – hopefully – that it doesn't need to be that long. Also, it isn't yet known whether active stretches, where you get into the position using your body's own strength, work better than passive stretching, where you, or someone else, adds extra force at the limits of your natural range. The animal studies hint that active stretching is better for reducing inflammation, but that will need to be confirmed. Answers should start to emerge over the next few years.

Moving along

In the meantime, it's worth looking at a separate but related benefit of stretching that could be just as important as the cellular changes. It may help to physically swoosh the fluid in the fascia along, allowing the body to give it a regular spring clean.

In 2018 a team of researchers at New York University led by pathologist Neil Theise was able to get a good look at a sample of human fascia in situ for the first time, using a new kind of microscope that can be attached to a small medical probe that is being used to take tissue for biopsy. Using this new technology, they were able to see that what had looked like a dense mesh of collagen fibres when it was removed from the body and squashed onto a microscope slide was actually more like a fluid-soaked sponge when seen in its natural state. Theise's studies suggest that, when the sponge gets squeezed, fluid drains into the lymphatic system, the system of pipes that recycles fluid from the tissues and passes it through the lymph nodes so that the immune system can check it for problems.

Having looked at samples of connective tissue from the gut, lungs, fascia and fat layers, Theise and his research team concluded that this sponge-like structure is a general feature of loose connective tissue throughout the body. It also seems to contain a surprisingly large portion of our body fluid. It has long been known that the fluid matrix that surrounds individual cells drains into the lymphatic system, where it gets cleaned up and recycled, but it was news that the connective tissue is part of this process too. This opens up the possibility of a body-wide network of fluid that allows crosstalk between different kinds of tissue and the immune system, with what the research paper calls 'compressible and distensible sinuses through which interstitial fluid flows around the body'.[11]

Theise estimates that this makes up 'approximately 20 per cent of the fluid volume of the body, comprising approximately 10 litres'.[12] This would make it not only the main source of lymph but also one of the body's major fluid compartments – the others being blood, the fluid inside our cells and the liquid padding around the brain and spinal cord.

The discovery was reported by many media outlets as a 'new organ' that had been hiding in plain sight. That in itself might be stretching things a little, but it was certainly an important find, especially when considering why movement is good for our health.

It may be no accident that loose fascia, with its gloopy, liquid matrix, tends to be found in parts of the body where movement is an integral part of the system. The gut, for example, slowly squeezes food from one end of the body to the other via a wave of muscular contractions that, presumably, also squeezes the surrounding connective tissue. Similarly, the lungs, bladder and heart expand and contract repeatedly day and night, flexing and contracting the tissues around and within them and squeezing the fluids along.

While the liquid layer surrounding these particular organs gets moved around by default, the fascia around some of our other organs, muscles and throughout the body cavity only gets moved along if we voluntarily move our bodies. 'Clearly fluid movement has important physiological results and such movement is provided in the musculoskeletal fascia by body movement,' says Theise. This would suggest that being sedentary is not the best way to keep the body's fluids flowing freely and to let the body deal with immune threats as and when they arise.

This reminds me of something that I've been told to do many times in yoga classes: wring the toxins out of your muscles. I've always dismissed the idea – frankly it sounds like one of those things that people believe because it sounds good, and maybe because it feels like that's what's happening when you get into a good, deep twist. But if movement is required to drain body fluids into the one of the systems that naturally detoxifies the body, then the idea of wringing out your fascia like a sponge starts to sound a little less far-fetched.

The true proof of the pudding would be if a study were to show that stretching and moving in general increase the throughput of fluid through the fascia and into the lymph. This we don't know because, to my knowledge, no one has done the studies, at least in healthy people. There have been some studies in cancer patients showing that exercise, including stretching, can reduce accumulation of fluid in limbs after part of the lymphatic system has been damaged or removed during treatment,[13] so it's possible that stretching and compressing the fascia that surrounds our muscles and organs moves things along, potentially helping the body deal with any problems that arise rather than leaving them festering in the tissues.

This, in fact, is very relevant to another piece of yoga

folklore – that the poses somehow clean out your organs. In the summer of 2019 I had the chance to ask a top yogi what this vague but intriguing nugget is really getting at.

Sharath Jois is the paramaguru (the 'guru's guru' or lineage holder) of Ashtanga yoga, a title he inherited from his grandfather, K. Pattabhi Jois, who died in 2009. Now in his late forties, the younger Jois is slim, strong and totally unassuming – most people would walk past him in the street without giving him a second glance. To serious Ashtanga yogis, though, he is a very big deal: somewhere between a deity, royalty and a Hollywood star. Many devotees practise in front of an altar bearing his image, and when I told a yoga teacher friend that I had arranged an interview with him during his visit to London, she asked if I was planning to kiss his feet (I wasn't).

We sat down for a chat at the dining table of the small London flat that he was using as a base while he was in town. As we got chatting about how yoga affects body and mind, it struck me that he didn't once mention the word 'stretch'. We Westerners often think of yoga as a way to stretch out tense muscles while improving flexibility and getting stronger. But Jois didn't even mention flexibility as a goal. He actually laughed when I asked whether the point of the postures was to stretch our bodies to make us less hunched and more human-shaped.

'No, it's not that!' he said. 'Actually, that also happens but […] what you are doing is you are exercising your internal organs so that they can function well […] when the organs are not functioning properly this causes health issues.'

Which seems to be close to what the scientists are saying. If that's the case, then getting your nose to your knees is not the point of yoga after all. Bendiness is a means to a more important end. And while I used to think the idea of massaging your internal organs sounded ridiculous (if they needed a massage, why would the body go to such lengths to put them in bony cages?), it could be that movement reinstates normal levels of fluid movement in the fascia surrounding the organs to keep things ticking along in a way that sitting hunched over a desk doesn't.

The good news is that you don't necessarily need to get your leg behind your head to get the benefits of stretching. 'It has to be gentle, I cannot emphasise that enough,' says Helene Langevin, back in Maryland. 'The amount of weight that we put on the tissues in the animals is measured in grams: it's very small. I would say less is better. I always think, one cell at a time. Just very tiny stretching, be respectful of the tissues – don't yank 'em. Slowly and gently, that's the key.'

And, as Jois points out via a helpful analogy, testing your flexibility to destruction isn't a particularly sensible goal. 'Your skull is strong – that doesn't mean you have to go and hit your head on a rock! That sense you should have.'

Too much of a good thing

Anyway, you can definitely be too stretchy, which can bring problems of its own for both body and mind. Around 20 per cent of people are hypermobile, meaning that their joints extend beyond the usual range of motion – a phenomenon more commonly known as being 'double-jointed'.[14] It's caused by an unusually stretchy form of collagen in the body's connective tissues, and while it's useful for ballet dancers, gymnasts and musicians to be able to stretch their joints to extremes, too much flexibility can also lead to chronic pain, joint dislocations and digestive problems like irritable bowel syndrome. More surprisingly, perhaps, it seems to be linked to mind-based symptoms too.

The observation that having bendy joints affects the way people feel dates back to 1957, when the Spanish rheumatologist Jaume Rotés-Querol recorded what he considered to be unusually high levels of 'nervous tension' in people with hypermobile joints. The observation was largely ignored until 1988, when a group of researchers at the Hospital del Mar in Barcelona also noticed that hypermobile patients seemed to be prone to anxiety. They decided to study the link in more detail. Since then, it has been fairly well established that there is indeed a strong connection between the two. One study found that 70 per cent of hypermobile patients had an anxiety disorder of some kind, compared with 22 per cent of healthy volunteers, while another estimated that anxiety and panic disorders are sixteen times more common in people who have hypermobile joints.[15] There is also an emerging link between hypermobility and eating disorders, chronic pain, fatigue and neurodevelopmental disorders, including ADHD and autism.

These problems may occur because having joints that keep flexing when you expect them to stop interferes with the process of interoception, the ability to sense the body's internal state, making it hard to pinpoint where internal bodily messages are coming from. Studies in Hugo Critchley's lab at the University of Sussex have provided some evidence for this: people with hypermobile joints were found to be unusually sensitive to interoceptive signals coming from within their bodies, such as heart rate and other stress-related changes. This sounds like a good thing, except for the fact that they also tend to be less accurate at identifying where in their body these signals are coming from, and less able to interpret what they mean.

Given the fuzzy nature of bodily signals in people with hypermobile joints, a racing heart can all too easily get interpreted as anxiety, which is particularly confusing when there is no obvious external need to be worried. Another study seemed to back this up, finding that the more sensitive a person is to their internal bodily signals, the stronger the relationship was between hypermobility and anxiety.[16]

Another problem with unusually loose collagen is that it can lead directly to an overactive fight or flight response, Critchley told me. It comes back to the fact that connective tissue is everywhere, and that its basic structure is pretty much the same wherever it is found in the body. That means that stretchy joint collagen often equals stretchy collagen everywhere else, including the lining of blood vessels. Under normal circumstances, when someone stands up from sitting or lying down, their veins automatically constrict to stop the blood pooling in the legs and causing a temporary dip in blood pressure. If the collagen in those vessels is too stretchy, though, the reflex doesn't work as effectively, blood pressure drops whenever the person is upright and the heart has to compensate by pumping harder.

The combination of over-sensitive internal signals and less than stellar control over the functioning of the autonomic nervous system could explain why hypermobility makes anxiety vastly more likely, even when nothing scary is happening. Given that a fifth of the population has hypermobile joints, this could be contributing quite significantly to the number of people who suffer from anxiety and are at a loss to understand why.

The link from hypermobility to ADHD and autism is less clear, but the Sussex team has some ideas. Studies have shown that people with hypermobile joints tend to be more sensitive to external sensory signals and pain. Perhaps, then, this combines with hyperactive internal signals in a way that predisposes people not only to withdraw from the outside world but also to dissociate from their own overwhelming feelings. Jessica Eccles, another member of the University of Sussex team, says that this is 'just conjecture' at this point, but there are early indications that it may be the case in other hypermobility-related disorders, such as fibromyalgia.[17] Whether the same can be said for autism and ADHD remains to be seen.

Eccles herself has hypermobile joints and is well aware that not everyone who is especially bendy is comfortable with being considered to be 'defective' in some way. Nevertheless, she says, there is a value in understanding these mind–body links, not least because they could bring comfort to people who have had their 'psychosomatic' disorders dismissed and been made to feel as if they are making them up.

It could also bring new treatments. There's not much that can be done about the physical make-up of the collagen itself – if you happen to have stretchy collagen, you're stuck with it – but knowing that this physical trait affects mental health via specific pathways opens up the possibility of using body-based interventions to change the mind.

One avenue could be to strengthen the muscles around the joints, not only to prevent overextension and reduce joint pain but also to put more rigid boundaries around the sense of self that is constantly being built and updated by the extent of the body's movements. Building stronger muscles, particularly in the lower body, may also help to reduce the heart rate rise on standing, by squeezing blood back upwards, potentially reducing both a racing heart and anxiety.

Another idea is to improve a person's ability to make sense of their own internal sensations. Interoception is a skill that can be improved with training – which is why Eccles is working on a study to see if helping people to pinpoint and understand their physical sensations will control anxiety in people with hypermobility.

In ADHD and autism these kinds of intervention could be used in childhood to reduce the chances of developing the anxiety and sensory processing problems that commonly come with both conditions. One approach, which is starting to be used in occupational therapy, involves playing games where children point to different body parts and say how they feel right now, with a teacher helping them to put a name to that emotion.[18] The idea is that this could help children learn to decode their own body signals more effectively and to begin to regulate these feelings, potentially reducing at least some of the suffering early and before they become too entrenched in biology. One of Eccles's early studies showed that people with hypermobility had larger than average amygdalae (the brain regions involved in emotional processing and best known for their role in fear) and smaller brain regions linked to the representation of the body in space. Teaching children with neurodevelopmental issues and hypermobility to understand their own moving bodies might prevent both of these things from becoming set in both body and brain from an early age.

A little of what you fancy

Given the problems that are attached to extreme flexibility, it would seem to follow that stretching for the sake of getting more flexible might not be a great idea for body or for mind. And since everyone's joints, muscles and connective tissue are different, there is no one-size-fits-all answer to how much we should stretch – or whether anyone who can actively move the joints through a healthy range of motion and has decent core and joint strength actually needs to get bendier.

This may sound strange, given the emerging links between stretching and immune function. But it's worth remembering that the kind of stretches that Langevin is recommending are not about pushing hard every day until you can finally touch your toes. She herself strives not for greater bendiness but for the pure enjoyment of feeling her body stretch. 'I don't do fancy yoga or anything. I stretch where I feel I need to stretch. And I love it – it's wonderful,' she says.

In the end, one thing we know for sure is that stretching after a period of immobility feels good. A nice bit of pandiculation when you get up from a chair is pleasurable, reminds your brain that you have some limbs after all and may help to swoosh the body's fluid along a little. It may even reconnect body and mind so that they function in the way that they should. But we are talking about moving and gentle stretching in the normal range of motion. Getting into the splits is impressive, but taking the hip joint back further than around 30 degrees from centre is surplus to requirements for any normal human movement.

Speaking as someone in the throes of a decade-long love affair with yoga and with some hard-won flexibility gains to show for it, this is hard to write and even harder to take. The founder and late guru of Ashtanga yoga, K. Pattabhi Jois (Sharath Jois's grandfather), is often quoted as having said 'body is not stiff, mind is stiff', something that is parroted by many a yoga teacher as they yank you into a ridiculous bind. There is actually some evidence to support this: while stretching does increase flexibility, it's not necessarily because it makes the muscles physically any longer. Instead, the nervous system is being re-educated that it's safe to take the joint beyond its current range. The body puts the brakes on when we get near the edges, to prevent injury. By gently going beyond this range, research suggests, you can persuade your nervous system to release a little bit more of the slack. The 'edge' of a stretch isn't where your muscle thinks it is but where your nervous system draws the line.

Even so, there's an argument for staying inside these lines, rather than persuading the joints to go too far outside their comfort zone. When I met Sharath Jois, I put it to him that it could be dangerous to take his grandfather's words too literally. He agreed. 'Certain times [...] your body is flexible but your mind says don't do it. So certain times you need to push yourself a little bit,' he said. But, he added, 'You should know your limitations.'

How to move: stretch

- **Pandiculate**: After a spell of sitting, stand up and stretch your arms and legs. It reminds your brain you have limbs as well as releasing tight muscles. Do it at least once every hour, if not more.
- **Move, stretch, twist**: To squeeze the fascia surrounding the muscles and organs and, potentially, keep the fluids

of the immune system moving along. Go gently, and don't aim to go beyond your normal range of motion; just move until you feel the stretch.

- **Strength before bendiness:** Combine gentle stretches with strength work, especially if you are hypermobile. Strength and flexibility together make for powerful weapons against anxiety.

7

Breathless

Regulate [the] breathing, and thereby control the mind.
B. K. S. Iyengar[1]

There's a moment in the film *Rise of the Planet of the Apes* (2011) when Caesar the chimp speaks for the first time, yelling 'No' at his abusive keeper before knocking him out cold. It's an eerie moment – and not only because it's an animal using language to make his feelings about our species perfectly clear. Even more unnerving is the way Caesar steadies himself afterwards by taking a few deep breaths.

There is a good reason why it is so disturbing to watch. Even if you've never given it a moment's thought, breath control is instinctively recognisable as not only a human-only skill but also as one that is intimately linked to our unique powers of mental and emotional self-regulation. Somewhere in the depths of our psyche we know that if other animals were to develop these skills too, combined with their far superior strength and agility, we'd be toast.

Thankfully, in the real world none of our closest relatives has shown anything like a human level of breath control.[2] But you could definitely make the case that our species isn't making anything like enough of this ability. Despite centuries of reports from followers of Eastern traditions that slow breathing can improve focus, bring a sense of calm when we might otherwise lose it and even whisk us away to an altered state of consciousness, most of us still don't take time out from our busy lives to prioritise this simplest and most unobtrusive of body movements.

Obviously we breathe automatically day and night, and most of the time we don't give it a second thought. And for a good while neither did the majority of scientists. Along with other air-breathing creatures, the brainstem, one of the oldest parts of the brain, has long been known to be in charge of setting the rate of respiration, making sure that we keep oxygen flowing into our blood via the lungs and that we let waste carbon dioxide out. It does this day and night, from the first seconds of life until the very last. In, out, in, out.

The precise cluster of neurons that makes this happen was discovered in the 1970s by a post-doctoral student called Jack Feldman. As any self-respecting student would, he impulsively named it after a bottle of German wine that happened to be on the table at the conference where he announced his discovery. Feldman went on to become a world expert in the region he named the Bötzinger Complex and has shown that, along with its neighbour, the pre-Bötzinger Complex (or pre-Böt-C), it is crucial for setting the rate and rhythm of breathing, and for ramping things up when there is a shortage of oxygen in the blood.

An even smaller cluster of neurons is in charge of making us sigh every five minutes or so. Recent work in Feldman's lab at UCLA suggests that this is a physiological reflex that stops the air sacs in our lungs from collapsing and sticking together like a deflated balloon.[3] Other species – dogs, mice, cats and so on – all sigh at slightly different rates, but for the same reason. So, while it might seem as if a dog is sighing because he would rather be out chasing squirrels but lacks the opposable thumbs to open the front door, it's much more likely that his lungs are automatically re-inflating after a period of shallow breathing.

Yet we do sigh to express emotions, including exasperation, sadness and relief. Psychology studies suggest that, as well as being a form of communication, emotional sighs – of the kind that people produce when psychologists give them puzzles to solve and don't tell them that they are unsolvable – act as a kind of 'reset' of the breathing system. Sighing returns us to normal after a period of stress-related shallow or irregular breathing.[4] Making a conscious decision to take control of the sigh reflex is the easiest way to deliberately control breathing for mental gain. A strategically timed deep sigh can act like a mental full stop and capital letter, making it easier to put a period of stress behind you before you move on to something else.

This ability to hijack the sigh reflex is entry-level breath control, however. The real power of breathing comes from the way that we can also control the rate and depth of our breathing rate and choose from a menu of body–mind benefits. Like Caesar the chimp, we can use it to calm down, focus and think about what to do next. With a little practice we can also use breath control techniques to escape from reality for a while, taking a well-earned break from both body and mind. Or we can make like a monk and admire the rhythmic work of the pre-Böt-C from a mental and emotional distance. All are easy to do and can make a significant difference to how you think and feel – for solid physiological reasons.

Move!

As a fidget myself, I know all too well that sitting around and tuning in to your breathing doesn't come naturally to everyone. For some, the New Age, meditative connotations of breath control are a deal-breaker. For others, sitting still on a cushion is time that could be spent doing something less boring instead. Whatever the excuse, there is now enough robust science to show that mastering the simple movements needed to control the rate, depth and route by which you get air into your body can become a handy tool to steer thoughts and feelings in useful ways. Mastering this range of bodily movements can allow us to dial into the workings of the brain and the rest of the body, change the settings of both and get the very best out of the mind, no matter whether you are on the move or tied to a desk.

No one really knows exactly why our species lucked out and got the skills to deliberately control the muscles that affect the way we breathe, but it's probably no coincidence that – as in the fictional case of Caesar the chimp – it appeared at about the same time that we became able to speak. Speaking, as opposed to grunting, moaning or yelling, requires the ability to control a long exhale, punctuated with a few perfectly timed inhales and some nifty control over the larynx, lips and tongue. Studies comparing the skeletons of different species of ancient humans who lived between 100,000 years ago and 1.6 million years ago showed that modern humans and, interestingly, Neanderthals evolved spinal columns with significantly more space for the nerves that supply the breathing and facial muscles than older species. This would have allowed them the kind of hardware necessary for fine control of the breath and the noises that it produced. While their predecessors carried on grunting at each other across the plains, this broader repertoire of noises developed into a useful new way to communicate.[5]

However we got here, we have ended up with a powerful tool. You don't have to be fit or flexible, strong or even particularly mobile to take advantage of its benefits: it's surprisingly easy. And, let's face it, modern life offers many situations where it would be an advantage to take control of your physiological, mental and emotional state.

Synchronise

An example: I'm writing these words on a brick of a laptop that is older than my pre-teen son and almost as heavy. Yesterday I spilled a full cup of tea on my own trusty machine, and it hasn't shown any signs of life since. It's not the end of the world – I probably lost a couple of thousand words and a few tweaks here and there – but, with only a few weeks and 20,000 words standing between me and my deadline for this book as I write, it would be fair to say I'm quite stressed. Looking on the bright side, though, this gives me the perfect opportunity to use one of the most useful aspects of breath control: it can provide a sense of calm focus, even when your head is spinning out of control.

In a bid to stop freaking out and get back to work, I found a seven-minute focus-based meditation on YouTube, sat up straight in my chair and did exactly what I was told. And, despite the urge to skip to the end, after a few minutes of focusing on slow, deep breaths my head stopped spinning, and the urge to scream, cry or lie down and die gradually faded away.

So, yes, it does work. But then I knew it would, because

as well as the experiences of several generations of monks, there is also an ever-increasing pile of scientific research that demonstrates that, when you control your breath, what you are actually doing is taking your brainwaves in hand and tethering the rate of their fluctuations to your breathing rate.

Brainwaves (also known as neuronal oscillations) are rhythmical pulses of electrical activity across groups of neurons as they send messages across the brain. When enough neurons fire at the same time, these pulses are strong enough to be measured via electrodes on the scalp and can be translated into graphs showing the peaks and troughs of activity. Since the technology to do this was invented around a hundred years ago, scientists have known that brainwaves come in a different range of frequencies, and that the dominance of certain frequencies at different times provides clues about the kind of processing that is going on (see table opposite).

As we saw with the entrainment of brainwaves to a beat, synchronisation of brainwaves across far-flung parts of the brain allows regions that specialise in different kinds of processing to pulse to the same rhythm, allowing different kinds of information – what we can see, hear and smell, for example – to be bound together as part of the same experience. This allows the brain to take 2 and 2 and make 5, by taking different kinds of inputs, putting that information together and making sense out of what it all means.

The link between this process and breathing comes via sensory neurons at the top of the nose. These neurons have two roles: they pass on information about the scents being brought in on the air to the olfactory bulb, and also detect the physical movement of air as it wafts past. Because of this dual

Frequency band	Frequency range (Hz)	Type of thinking
Gamma	>35	Problem-solving, concentration
Beta	12–35	Busy mind, externally focused attention, anxiety
Alpha	8–12	Relaxed, reflective, passive attention
Theta	4–8	Deep relaxation, drowsy. Internal focus
Delta	0.5–4	Sleep

Source: *Introduction to EEG- and Speech-Based Emotion Recognition*, 2016, pp. 19–50.

purpose, the regular in–out movement of the breath in the nose acts as a metronome that sets the timing for scent-based information to hit the brain. Given that this information says a lot about how safe, or rewarding, the environment may be, it makes sense in an evolutionary context that other salient information, perhaps from memory, would hop onto the same rhythmic frequency band.

Animal studies suggest that this is exactly what happens. Synchronisation between breath and brainwaves happens first in the olfactory bulb, where scents are detected, but then spreads further, to brain regions that deal with assigning meaning to the scent. Research in mice has shown that the breath-related rhythm spreads to areas of the brain that deal with memory – allowing the animal to decide whether this particular smell is something that has been encountered before – and to the emotional centres that can help determine how to respond.

A study from 2016 led by Christina Zelano, a neurophysiologist from Northwestern University, in Chicago, was the first to confirm not only that the same phenomenon happens in humans but also that in our brains the synchronising effect stretches even further, bringing the thinking, planning and decision-making areas of the prefrontal cortex along for the ride. Some researchers reckon that having brainwaves locked to the rate of breathing is a general feature of the way the brain works.[6]

Inspiration

Studies of brainwaves during breathing suggested that the strongest effect of synching up with breathing rate comes on an in-breath. It sounds a bit cheesy, but it's also true: when we breathe, we are literally taking inspiration from the environment and the subtle clues that it contains. In fact, that's almost exactly what yogis and martial artists do say: in martial arts 'qi' (sometimes spelt 'chi') means breath but also focus and power. In yogic breathing, or pranayama, you're said to take in 'prana', which translates as breath, energy and universal consciousness. Sharath Jois, the current guru of Ashtanga yoga, tried his best to explain the benefits of this to my Western ears. 'When we breathe, it's like we are taking inside the positive energy through nature, from outside, through breath,' he said.

Personally, I'm more comfortable with a more scientific

way of saying the same thing: that, as well as oxygen, inhaling brings information about the world and a chance for our brainwaves to beat to a common rhythm, changing the way we feel.

And this is where breath–brain synchronisation becomes a useful tool to change your mental state. Research where people consciously change their breathing rates has shown that different ways of breathing can encourage particular frequencies to dominate across the brain, which can take us to a more alert and focused state or to a more relaxed and drowsy one.

There's only one catch, and it's something else that yogis have been banging on about for centuries: breath-linked mind control only works if you breathe through your nose. According to some estimates, more than half of people habitually breathe through their mouths. This not only contributes to bad breath and tooth decay: it also bypasses the nose-to-brain hotline.

The people we have to thank for the evidence behind these insights are a handful of epilepsy sufferers who volunteered to take part in research while in hospital having invasive brain surgery. In cases where the seizures originate in a particular region of the brain, and the available drug treatments have failed, patients can opt for surgery to remove the area of the brain that sets off their seizures. Pinpointing the part of the brain that starts off the erratic electrical activity of a seizure while leaving the brain regions that control speech, movement and other vital functions intact is a tricky business. It's done by removing a portion of the skull, putting electrodes onto or into the brain and waiting for a seizure to happen, recording the brain's electrical activity while the patient, and their surgeons, wait.

It can take several days for a seizure to hit, which leaves patients stranded in hospital, wide awake but wired to a monitor by their head and with nothing much to do. Happily, neuroscientists are more than willing to provide them with entertainment in the form of experiments that help to localise various brain functions to activity in certain areas. The vast majority of electrodes are implanted into healthy areas of the brain, which means that neuroscientists can record activity in healthy, awake human brains while they are doing various tasks and see exactly what is happening in individual neurons and across larger areas of the brain.

Using these methods in eight patients, Zelano confirmed that breathing does indeed act as a conductor of human brain activity, particularly in the processing of memory and emotion. The more closely that brainwaves synchronise with breath, the better people were able to store and retrieve information from memory, and the faster they were able to react to signs of danger. In the experiments Zelano also found that volunteers were significantly faster to react to an image of a frightened face if it was shown to them while they were breathing in.

Mind your nose

Nasal breathing, though, was the key. In Zelano's experiments, when the volunteers did the same tasks while breathing through their mouths, the synchronisation of breath to brainwaves was much reduced, and their reaction time to the emotionally charged faces was significantly slower.

It's important to point out that breathing through the

nose didn't provide any boost in terms of understanding what they had seen – people were just as accurate at spotting a fearful face whether they were breathing through the mouth or nose. Intriguingly, though, nasal breathing made them significantly faster at *moving their bodies* in response to what they had seen. In the experiments, this was a simple matter of moving their finger to press a button, and the difference was measured in milliseconds. Even so, in real life, that could make the difference between getting out of the way of a speeding truck and becoming a pile of raspberry jelly.

People were also more accurate on memory tests on an inhale (but again, only when they were breathing through the nose), suggesting that, while avoiding becoming raspberry jelly, you might retain the lesson of looking before you cross more effectively if you learn it on an inhale.[7]

This not only provides a good reason why we breathe faster in an emergency – to take in as much information as possible – but also suggests that, as long as you can stop panicking and remember to do it, consciously deciding to breathe in slowly and deeply in times of stress might help you see the wood from the trees and make better decisions. It also suggests that a session of deep breathing before an exam (or even while racking your brain to remember what was on your shopping list) might help dredge useful information up from the depths.

Calm down, tune in …

Another thing that yogis talk a lot about is the idea that breathing brings focus to the here and now, to put a wandering mind back with its body where it belongs.

'In yoga we call it *chitta vritti*,' Sharath Jois told me. '*Vritti* means it's going different directions ... So, what you can do through yoga – asana [posture] practice and breathing practice – we can bring that mind in our control.'

A study from 2018, with a different set of epilepsy patients from Zelano's, though they were also undergoing surgery, has shown that slow and deliberate breathing does indeed improve the ability to focus, while also increasing bodily awareness. In one set of experiments at North Shore University Hospital, New York, neuroscientist Jose Herrero teamed up with neurosurgeon Ashesh Mehta and asked eight patients simply to breathe normally and count their breaths, bringing awareness to their breath but not changing it in any way. Then, in a separate experiment, they asked them to concentrate on controlling their rate of breathing, by taking faster breaths than normal.

Between them, the eight patients had 800 electrodes inserted into thirty-one different brain regions, so the research team was able to map activity across a wider range of networks than studies that had gone before. Because of this they were able to track synchronised brain activity across different networks of the brain and to see whether it changed depending on the kind of breathing the people were doing.

They found that, when the volunteers passively observed their breath but didn't change its rate or rhythm, brainwaves in areas involved in interoception – the sense of the internal state of the body – became more strongly locked to the respiration rate. This could be important: as we've seen with dance and stretching, tuning in to how your body is feeling can be a powerful tool to understand and manage your emotions. The study volunteers only counted their breath for a couple of minutes at a time, which suggests that taking just a short break to do the same – no cushion or chanting necessary, you don't even need to close your eyes or let on that you're doing it – could be a great way to get out of your head and calmly reconnect with the rest of your body. Over time a regular check-in using breath as a short-cut to the body could add up to a significant boost to mental health.

Passively observing your breath without changing it is a key feature of mindfulness meditation, which may explain why so many studies suggest that it can improve both interoceptive ability and mental health. But even those who are diligently doing ten minutes of mindfulness meditation a day, passively following their breath without judgement, may still be missing out on some of the benefits of breath control. According to Herrero and Mehta's study, actively controlling it does something else entirely.

When people were asked deliberately to change the rate at which they were breathing, synchronised activity showed up in different parts of the brain, specifically in circuits that are known to be involved in sustaining attention and focus. Other studies suggest that the act of focusing on breath decreases theta waves – those that signify a zoned-out state – and increases alpha waves, which are associated with relaxed alertness.[8] This, according to studies of sustained attention, is the best possible state to stay focused for long periods.

... and drop out

Not only that, but deliberately changing the speed of your breathing metronome can have a big impact on how you feel.

Left to its own devices, the Pre-Böt-C will keep the rate of breathing between twelve and twenty breaths per minute at rest, but this rate can shoot up as high as thirty breaths per minute during the kind of hyperventilation that accompanies a panic attack.

Deep, slow breathing is a tried and tested method of bringing a panic attack under control, bringing oxygen and carbon dioxide back in balance and telling the body to move from fight-and-flight back to a normal level of arousal. If you are already breathing at a normal rate, slowing it down further still can change your state of mind to the point where you tune out from reality and skip happily away with the fairies.

Buddhist monks have mastered the art of breathing at three or four breaths per minute – taking one breath, in and out, over the space of twenty seconds. This kind of slow, controlled breathing never happens by accident and can only come as a result of making a conscious decision to override the status quo. It's not easy, but it can be done and, according to a recent study, it's worth the effort, particularly if you are looking for a drug-free route to an altered state of consciousness.

Andrea Zaccaro, of the University of Pisa in Italy, found himself intrigued by tales of monks reaching an alternative plane of existence where they feel at one with the world and everyone in it. He wanted to find out whether it was slow breathing that produced this marvellous result or whether it is a side-effect of mentally zooming in on the breath to the exclusion of everything else. In other words, are the mental changes the result of physiological side-effects of deliberate breath control or are they purely a brain-based phenomenon that is a side-effect of the act of focusing, and which has little to do with what was happening from the neck down?

To find out, Zaccaro recruited fifteen student volunteers, this time with their skulls left intact, and strapped electrodes to the outside of their heads. Then, using the kind of nasal cannula you see in hospitals when patients require additional oxygen, the research team blew air up their nostrils to simulate breathing at a rate of three breaths per minute, for fifteen minutes. Cannula aside, the nostrils were plugged to prevent normal nasal breathing, but people were still allowed to breathe freely through the mouth. It sounds incredibly uncomfortable but apparently wasn't. In fact, two volunteers had to be excluded from the study analysis because according to their brainwave traces they fell asleep.

The volunteers who managed to keep their eyes open showed synchronisation of brainwaves across the brain, this time at the low frequencies of delta and theta waves. These frequencies were particularly strong in areas involved in emotional processing and what's called the default mode network (DMN), a group of brain regions that acts as a kind of neutral gear that the brain slips into when it is not focused on any particular task. The DMN is also active when we are having internal thoughts about ourselves. Theta waves are those that accompany deep relaxation, a feeling of being mentally detached and focused on the internal state rather than the outside world. As you would expect from this kind of brain activity, volunteers reported feeling deeply relaxed and content during the experiment, and many reported being outside of their own minds, in a state of 'being' rather than 'thinking'.

This lovely-sounding feeling could explain what keeps

expert meditators so committed to all that sitting around. The power of three breaths per minute could provide a much-needed holiday from thinking and a freeing sense of being part of something greater than ourselves.

Whether you believe that this is evidence of a greater spiritual power or a global consciousness that we can connect with, given enough dedication, or whether you think of it as a biological phenomenon that makes you feel good, the important thing here is that, thanks to the way that your brainwaves synchronise with the rate at which you breathe, anyone can access this feeling for free. All you have to do is take control of your diaphragm and the intercostal muscles between your ribs, and practise slowing it down to three breaths per minute.

The magic of six

Breathing at the rate of three breaths a minute does take a bit of practice, though. And, as Zaccaro's study volunteers found, it's not easy to stay awake for long enough to get the feeling of being at one with the universe.

Six breaths per minute is a lot more manageable, and, according to research, seems to be even better for our physical, mental and emotional health. Breathing in and out over the space of ten seconds hits a physiological sweet spot that connects the breathing-related movements of the body to blood flow, blood pressure and the concentration of oxygen in the blood. Plus, it tips the balance of the autonomic nervous system from 'rev up' to 'calm down'. Breathing at six breaths per minute is, to all intents and purposes, a short-cut to a sense of calm and contentment.

This probably explains why it also feels really good. In experiments where volunteers were asked to breathe at different rates and report back on how it felt, people said that six breaths per minute felt the most comfortable and relaxing. At some level, humankind seems to have intuitively worked this out a long time ago. A study from 2001 found that ancient spiritual practices, from reciting the rosary (in Latin) to chanting yogic mantras, had a side-effect of slowing the breathing rate to six breaths per minute. This, the researchers speculated, may account for the sense of calm comfort that these practices bring to their many devotees.[9]

Even if you don't believe in anything, from a spiritual point of view, breathing at six breaths per minute will calm even the most committed atheist. You don't even need to consciously count your breaths, let alone memorise a chant or prayer. Simply doing diaphragmatic breathing – also known as 'belly breathing', is enough to make this happen all on its own. For the benefit of beginners, the easiest way to do it is to lie on your back with your knees up, one hand on your chest and the other on your belly. Then slowly inhale until you feel your belly rise, and allow your ribs to expand out and down. Your chest shouldn't rise much, if at all. Then, at the top of the breath, engage the stomach muscles, pushing the belly down, which sends the diaphragm back up and the air rushing out of your nose. With practice you should be able to do it sitting up and maybe even while moving around.

The most direct way that this can affect the mind is simply by getting more oxygen into the bloodstream. At six breaths per minute the lungs recruit and fill the largest proportion of alveoli – the air-filled sacs where oxygen diffuses into the blood while carbon dioxide filters out – which makes it the most efficient breathing rate to get oxygen into the body. Part of the explanation for this is that you can't breathe at six breaths per minute without actively pushing the breath out of your body. In automatic breathing this doesn't happen – air leaves the lungs passively when we stop actively expanding the lungs, let go of the diaphragm and let the ribcage fall back into place. Actively pushing air out empties the lungs more fully and this, studies suggest, leaves a larger vacuum for new air to rush in and fill. This, in turn, significantly reduces the amount of dead space in the lungs – the proportion of air that is essentially wasted because, while it enters the body, it never gets as far as the alveoli before being breathed back out.

Because of these mechanisms, deep breathing can increase oxygen saturation in the blood to the tune of a couple of per cent, which may be enough to make a small difference to our ability to think clearly.[10] In experiments where people were given cognitive tasks to do with or without being given extra oxygen to breathe, people who received an oxygen boost did slightly better.[11] Measurements of the oxygen content of volunteers' blood revealed that breathing in air with added oxygen boosted oxygen saturation levels in their blood by a couple of per cent – around the same boost in oxygenation that has been seen when people breathe at six breaths per minute. No one has yet done the experiments to confirm that deep breathing itself improves cognitive ability, but if it increases oxygen saturation and a similar oxygen increase by artificial means improves cognition, then it's not a huge leap to think that it probably does.

It's worth pointing out, though, that even the most shallow-breathing, sedentary person in the world is not in any danger of suffocation. Oxygen saturation levels in the blood range between 96 and 98 per cent, and the body is pretty good at keeping it within these bounds. It is possible, though, that getting a little more now and again could provide a temporary boost to both alertness and performance. Oxygen, along with glucose, is a staple of brain function: at least up to a point, more is better.

Vagus, baby

Alertness aside, the other benefit of slow-deep breathing is a feeling of intense physical and mental relaxation. This comes via a separate, but connected, mind–body pathway that is also tuned to the magic rate of six breaths per minute.

This link comes via the vagus nerve, one of the longest nerves in the body, which originates in the medulla in the brainstem (the same region where the Bötzinger Complex is found). From here it runs all the way to the end of the digestive tract, stopping en route to check in with the heart, lungs and gut. If you could see inside your body, the vagus nerve would look like two long bits of string, about the thickness of garden twine, one running down each side of the neck and then branching off into several thinner sections that make contact with the organs.

The vagus nerve is long, thick and obvious enough to have been spotted by the earliest anatomists. The first written account comes from the Roman anatomist Galen of Pergamon in the second century AD. At the time, very little was known about the way the human body worked – Galen's other notable discoveries include the fact that arteries carry blood, rather than air. If only Galen had worked out the power of this long and winding nerve, humanity could have saved itself a huge amount of stress, because we now know that it serves as a vital conduit that relays information back and forth to the brain, with updates about what is going on in the body and information about how we should think, act and feel. It also regulates inflammation, the immune-based mind–body phenomenon that, as we saw in the previous chapter, we'd all do well to get under control.

Around 80 per cent of the fibres that make up the vagus nerve run from the organs of the body back towards the brain, where they keep the chatroom up to date with the latest news. The other 20 per cent or so run in the opposite direction as part of the parasympathetic nervous system, which specialises in keeping the body in a state of relaxed calm when there is nothing to worry about. Believe it or not, this is supposed to be the body's default state: it may sound like a distant dream, but you are meant to be relaxed and calm unless there is something truly important or life-threatening to worry about. And even when something big and scary does come along, the job of the vagus is to put the body in a state of 'rest and digest' as soon as the need to 'fight or flee' has passed.

Messages are constantly heading in both directions, meaning that when you feel calm, vagus activity is high and breathing rate, heart rate and blood pressure all slow down. It is also true that if you can find a way to slow your heart rate and reduce blood pressure you will start to feel more relaxed. The good news is that the two-way nature of the communication channel makes it an easy system to hack – and breath control is the key that unlocks changes in all the others.

Remarkably, it's possible to use your breath to train your body to react more healthily to stress, both in the moment and over the longer term, by virtue of the way that it changes the level of activity along the vagus nerve. Over time, practising slow breathing can change your baseline level of stress reactivity to a point where you freak out less often and recover more quickly when you do.

The level of baseline activity in the vagus nerve (called vagal tone) can be measured pretty easily, albeit indirectly, by tracking heart rate variability, a measure of the length of intervals between two consecutive heartbeats. It's something that is easy to do at home via various phone apps and most smart watches.

The details of why heart rate, vagal tone and breathing are so tightly integrated are fairly complex, but it essentially comes down to changes of pressure inside the chest as it rises and falls.[12] When we inhale, the diaphragm moves downwards and the ribcage expands, which together increases the space inside the chest cavity and reduces the pressure exerted on everything that's inside the chest. That includes the major arteries sending blood to the heart, which under reduced pressure expand, allowing more blood to flow through them. Stretch receptors inside the arteries detect this change and send a message via the vagus nerve that more blood is on its way and that it should take off its brake to allow the heart to pump faster.

When we breathe out, the opposite happens. The diaphragm rises, the chest pressure increases, squeezing the blood vessels running back to the heart and reducing the signal to the stretch receptors in the veins. These tell the vagus nerve that blood flow has dropped, and it should put the brakes back on the heart rate, slowing it down and perhaps saving energy by preventing unnecessary heartbeats while the available oxygen has been used up and the lungs are emptying.[13]

All this on-off activity means that the heart rate is constantly speeding up when we inhale and slowing down when we exhale. Variation in the timing of the heartbeats, then, provides a proxy for vagal tone. Ultimately, a more variable heart rate is better, because it means that the vagus nerve is kicking in with each breath to keep heart rate nice and low. If heart rate variability drops, it's a sign that the body is under stress and the vagus nerve has been temporarily put on the bench.

For reasons that are not fully understood, the biggest rise in heart rate variability (HRV) has been found to occur when people breathe at six breaths per minute. And the effects seem to last – in a study where people breathed for thirty minutes at six breaths per minute, HRV increased both in the moment and for a short time afterwards, and people were reportedly more likely to use body-based emotional regulation strategies afterwards, suggesting that their interoceptive pathways got a boost too.

There are any number of reasons why you might want to use this pathway to calm yourself as often as possible. High vagal tone (high HRV) has been linked to better working memory and ability to focus, emotional stability and reduced risk of anxiety and depression. People with high HRV also have a better ability to control blood sugar levels and are better able to suppress inflammation. This isn't because they have a lower stress response in general – it's because with higher vagal tone they are better able to turn it off again afterwards. The stress response is both healthy and necessary. Increasing vagal tone is not about never getting stressed, but having a flexible system that can return to normal as quickly as possible afterwards.

Breath in motion

Given the power that controlling breath can have on the mind, and the changes that we now know come as a result of moving the rest of the body, it makes sense to look at what happens when you put the two things together and move in time with the breath. Synchronised movement and breathing is the mainstay of many mind–body practices, from yoga, tai chi and qigong to exercise like swimming, running and cycling, in which breath often slips into the rhythm of body movements.

Surprisingly, given the huge amount of research into sitting meditation, very few studies have looked at whether moving meditations do anything different to our minds from sitting ones – or indeed whether there is any difference between consciously moving with your breath and any other kind of exercise.

One review of the handful of studies that have been done was published in 2013 by Peter Payne and Mardi Crane-Godreau of Dartmouth College's Geisel School of Medicine. Their answer to both of these questions was: maybe. Some studies have reported that qigong improved mood more than stretching alone and as much as talk therapy, while others found that mindful movement was a more powerful tool than conventional exercise in improving quality of life and self-efficacy.

Payne and Crane-Godreau added the fairly major caveat that most of the studies that have been done so far are poor quality and often don't feature a decent control group as a comparison. Even so, they point out, given that mindful movement tends to be far less strenuous that conventional exercise, the fact that it seems to affect the mind at least as strongly as other forms of exercise makes it worth a closer look.

Payne and Crane-Godreau also note that mindful movement tends to involve breathing at the magic rate of six breaths per minute. Given that this is the sweet spot at which breath and blood flow are in perfect synchrony, they speculate that it could explain the sensation of moving 'qi', or 'prana', both of which translate as both breath and 'energy', around the body. '[The] alteration in blood volume may be part of the basis for the suggestion [...] that one "breathes into the arms and legs" – an obvious physical impossibility, but perhaps a good description of the experience of a regular oscillation of blood volume,' they write.[14]

You don't have to worry about moving 'qi' or taking in 'prana' to get the benefits, though. Based on my own mini-experiments while out with the dog, breathing at six breaths per minute is quite doable when you are also walking at 120 steps per minute. This, you'll recall, is the speed that also seems to optimise blood flow to the brain by means of a perfectly timed foot massage. The easiest, practical way to put this into practice is to walk in time to one of the many songs that have their dominant rhythm at 120 beats per minute (examples include, 'Celebration' by Kool and the Gang, 'Just Dance' by Lady Gaga and 'Rumour Has It' by Adele, but there are loads to choose from – just Google your favourite music genre and 120 bpm). Then, as you stride along, breathe in for five steps and out for five. It's more of a march than a meander and I, personally, couldn't keep it up for more than a few minutes at a time, but it did work as a short-cut to a kind of energised focus that was a great way to reset after a morning of sitting.

This chimes with David Raichlen's idea that walking feels good because we were built to be 'cognitively engaged athletes'. Perhaps walking fast and breathing slow puts us into the right frame of mind to hunt and gather while also keeping us relaxed and open to the world around us, with a boost to memory and focus.

Whether you choose to breathe on the move or while sitting still, what's important is that breathing at six breaths a minute seems to be a sweet spot that links breath, body and mind in a way that can dramatically improve both physical and mental health. And if life gets too much, you can always slow it right down to three, taking a brief holiday from the world and everything in it.

How to move: breathing

- **Sigh after stress:** It resets the respiratory system after a period of shallow breathing, allowing you to take stock and move on.
- **Six breaths per minute:** Breathe in for five seconds, and out for five seconds. It not only maximises oxygen intake but also stimulates the vagus nerve, which is part of the parasympathetic nervous system, which calms the body.
- **Three breaths per minute:** Breathe in for ten seconds and out for ten seconds. It takes practice but can take you to an altered state of consciousness where you can just 'be'.

- **Breathe through your nose:** Nasal breathing synchronises brainwaves to the rhythm of the breath via sensory neurons at the top of the nose. This can enhance memory and focus and may even prime the body to move faster in an emergency.

8

And ... stop

If you get tired, learn to rest, not to quit.
Unknown[1]

Rest. It's the inevitable antidote to all that movement, and sooner or later we all have to give in to the urge to flop. This doesn't make it any less true that our minds – and bodies – work best when we are on the move, and that most people are not moving enough. But there's also an argument to make that we aren't resting properly either. Getting the right balance between activity and rest is an important part of how movement can improve your life, so it's worth briefly looking at what it means and how to do it.

Although almost everyone feels as if they are on their knees with exhaustion most of the time, very little research has been done into what rest actually is – waking rest as opposed to sleep. While sleep is definitely restful, rest and sleep are two very different things. The most obvious difference is that without sleep we die. Rats deprived of sleep die within weeks, and people with a rare genetic disorder that progressively robs them of sleep die within twelve to eighteen months of diagnosis.[2]

Slow-wave sleep, in particular – the deep stages when

people are difficult to wake – is particularly important for our health, being both crucial for memory processing and storage and also as the time of night when the brain gets an internal clean. The fluid that bathes the brain and the spinal cord washes through the brain, clearing away waste products that have accumulated during the day, including rogue proteins that are linked to Alzheimer's disease.[3] Dreams, which mostly happen during REM sleep, seem to have a role in processing emotions, which may explain why a lack of sleep leaves us not only fuzzy-headed but also cranky.

Sleep is also the time when the body takes the chance to rebuild. Release of growth hormone from the pituitary gland enhances growth and repair while the immune system uses the downtime to take stock, tweak the numbers of circulating immune cells and dampen down excess inflammation.[4]

Overall, sleep is an important partner to an active life on the move, keeping us mentally, emotionally and physically on top form. The current advice from the experts is to aim for at least seven hours per night,[5] to keep to regular sleep and wake times, and to avoid caffeine, screens and large meals before bedtime. Do that, and with luck the mental and physical benefits will appear all on their own.

Waking rest is almost as important but, unlike sleep, it is voluntary. It is also woefully underappreciated in Western culture. The cult of busy-ness has brought us to a point where resting is seen as a selfish indulgence. Yet reports of burnout are coming from all corners of society, from students to medical professionals to perfectionist parents. Clearly, despite the rise of the sedentary lifestyle, whatever we are doing during all those hours we spend sitting around, we are not finding it particularly restful.

Perhaps because rest is underappreciated in modern times, it has barely been studied as a route to better well-being. In an attempt to fill in some of the gaps, between 2014 and 2016 a group of scientists, artists and writers working with the Wellcome Trust undertook the largest survey to date on the subject. They asked 18,000 people from over 135 countries what rest meant to them, how much they felt they needed and how much they actually got. The results, published in 2016 as 'The Rest Test', showed that 60 per cent of respondents said that they didn't feel they got enough downtime in their lives.[6] And, as if to underscore the fact that we think of rest as being morally wrong, more than 30 per cent thought that they were unusual because they seemed to need more rest than others.

This is a problem, because a lack of rest plays havoc with our mental and emotional lives, sapping concentration and making us feel frazzled and emotional. 'The Rest Test' found that people who reported feeling the most rested had the highest scores for overall well-being.

So, how to balance the need for rest with the dangers of a sedentary life? The solution is to rest smarter, ensuring that it soothes and restores body and mind, powering us up to get moving again.

In her book *The Art of Rest*, Claudia Hammond lays out the basic recipe, based on the Wellcome survey and other relevant scientific research.[7] She found that – as with movement – there isn't one 'weird trick' that works for everybody, but there are some general rules to resting well.

One important point, and a key difference between rest and sleep, is that rest doesn't necessarily need to involve being physically still. Hiking up a hill can count as rest if it gives you a mental break in the moment and a feeling of satisfied exhaustion afterwards. Likewise gardening, playing an instrument, having sex or playing sport: rest can be as active as you like, as long as it takes your mind off your woes for a while and leaves you feeling relaxed and restored.

As for the perfect amount of rest per day – it seems that you can definitely have too much of a good thing. People who rated their well-being the highest were getting, on average, between five and six hours of rest per day (not necessarily in one sitting). Any more than that and it started tipping over into boredom and guilt, which are stressful. And any rest has to be voluntary – if you're resting because someone has told you to, it just doesn't work.

A particularly interesting finding was that almost all of the activities that were rated as restful were things that were done alone, including reading, walking and listening to music. As an introvert this is blatantly obvious to me, but apparently it was equally true for the extroverts in the survey. Psychologist Felicity Callard, who was a Rest Test panel member, speculated that people found alone-time restful because it allowed them to tune in with how they were feeling.

This is an important point, which brings us round full circle to one of the main benefits of moving more in the first place: that it allows you to tune in better to what is going on below the neck, and to put the mind back in the body where it belongs. Once mind and body are better connected, it is far more likely that you will recognise the body's rest signals and act on them.

In practice, it can be difficult to tell the difference between being physically exhausted and in need of rest or merely being a bit lethargic, in which case moving will probably help. And since many of us are also mildly sleep-deprived, sleepiness enters the mix too, making things even more complicated. It doesn't help that the body's fatigue signals feel physically identical to lethargy – and they often come together, so distinguishing between them involves a bit of internal detective work.

Common sense comes in here. If you've been sitting around for ages, even using a lot of mental energy, it's likely that lethargy has set in. Lethargy is more of a motivational issue, whereas fatigue is more a sign that the body has done enough and needs to recoup some energy. A sensible first step to deciding whether it would be better to physically shake off feelings of lethargy, or down tools for a while and rest, would be to take some time, alone, to find the much-needed mental space to check in with how you're feeling.

This sounds easy enough (obvious, even), if it weren't for the fact that inflammation throws a major spanner in the works when it comes to reading your body's signals. It serves as an important rest signal, telling us that the body has been damaged or infected and needs to prioritise any available energy for recovery. But, as we've already seen, inflammation also ramps up in times of stress, even when the danger is more in the mind than the body. This explains why mental stress can be so physically exhausting and leave us in no mood to dance, run or do any other kinds of physical activity. In these circumstances inflammation tricks us into feeling as though the body needs rest when, in fact, it needs the exact opposite.

To deal with this kind of stress-related fatigue there are two options, both of which involve movement. One is to do some high-intensity exercise. Vigorous physical activity briefly increases the levels of inflammatory markers in the blood, which sounds like a bad thing, but remember: inflammation is only a problem if it carries on unchecked. A brief peak gives the body a very clear signal that it now needs to douse the flames to bring matters back under control. You can think of it like a clear-out that prompts the body to sweep the decks of immune activity. Alternatively, less vigorous activity, such as a walk, some tai chi or yoga or just some sitting and breathing, can bring inflammation down by hacking into the stress response, turning it down and sending a message via the vagus nerve that all is well. However you prefer your stress-busting movement, it can take you from lethargic to healthily chilled in less time than it takes you to convince yourself you are too shattered to move.

All of this together suggests that the modern scourge of tiredness is partly explained by a lack of movement and partly by a lack of proper rest. Working on one without the other is only going to give you half the improvement in overall well-being. We need to move to be still, and only from that place of stillness can we move well.

How to: rest

- **Alone:** Take the time to tune in to your sense of interoception, to identify what kind of rest you need, whether mental, physical or both. Gentle, mindful movement, such as stretching or breathing will help.
- **Don't rest all day:** Any more than five or six hours becomes boring and stressful, according to research.

And ... stop

- **Move:** Rest doesn't need to involve being still. Being physically active is one of the best ways to allow a busy mind to rest. Go for broke and sweep the decks free of lethargy. Or take an easier pace and let your mind wander where it will.

SUMMARY

Move, Think, Feel

Nothing happens until something moves.
Albert Einstein (attrib.)

It's one thing to point to all the scientific evidence that connects body movements to well-being, quite another to have the proof of the pudding show up unexpectedly and slap you hard in the face. If I had ever doubted the link between movement and a healthy and focused mind, the strange few weeks of lockdown during the COVID-19 pandemic of 2020, which coincided with the latter stages of writing this book, certainly hammered it home. Never had physical activity, fresh air and exercise been so obviously connected to the way my family and I were feeling, and to how well we were able to concentrate when the trappings of normal life were flung out the window and the world got scary and strange.

It quickly became obvious, for instance, that attempting to home-school my son while also keeping to a writing target of 500 words a day was far less likely to end in door-slamming if we started the day with a Joe Wicks virtual PE session on YouTube, instead of scrolling on screens in our pyjamas. A bounce on the trampoline was a good way to punctuate the change from one school subject to the next, providing structure to the day, an outlet for nervous energy and frustrations and resetting our ability to focus. Having spent months – years – writing every day without so much as a single trampoline break, I'm gutted I didn't think of it before.

As the days dragged on, it slowly dawned on all of us that the days when we sank into a family-wide sense of gloom were the ones where we left our state-sanctioned hour of fresh-air-and-exercise until later in the day. Once we were finally on the move, no matter how much bickering and sulking had gone before, a long walk or bike ride together always, always got us talking and laughing by the end.

Lockdown was tough, but in many ways it was also a gift. It provided a microcosm of the mental and emotional price of a sedentary life and a clear demonstration of the relief that a well-timed bit of movement can bring. When you're only allowed out once a day, as was the rule in the UK, the before and after is so much more obvious.

The people I met while researching this book didn't need quarantine to teach them this – they were onto it a long time ago. My aim in writing this book was to combine their experiences with a growing scientific understanding of what movement does to the mind, to convince myself – and everyone else – to put the body and the way we move it at the centre of our quest for health, happiness and overall well-being.

Thus far I've tried to do this by breaking down the question of what movement does to the mind into various component parts and looking at them in isolation. And although this is the exact thing that holistic practitioners have been criticising Western medicine for doing for years,

I make no apologies for doing so – it's a crucial first step in working out what is going on in there. For it to be any use in daily life, though, it's essential to put all of these moving parts back together to come up with some usable suggestions about how what we know right now might be put into practice.

A sensible first step is to look at some of the common themes that kept recurring in my conversations with scientists and expert movers. And there were many – at times it felt as if people were trying to do exactly the same thing through very different methods. In many ways they were – because many forms of mind-altering movement tend to tap into the same key body–mind hotlines. Hit all of these spots and you can move however you like and reap the same benefits.

To summarise, the must-have elements of any movement plan include:

1) Defying gravity

Forget fancy gym equipment: human bodies were built to work against the downward pull of gravity. Putting weight on your bones and moving stimulates the release of osteocalcin from the bones, which boosts memory and overall cognition and may reduce anxiety.

Moving – and resting – in a way that doesn't involve collapsing in a heap (kneeling, squatting, sitting upright without leaning) also keeps the core working – with potential benefits for a strong stress response as well as slightly less spongy abs. And when working against gravity to move, compression on the soles of the feet helps blood flow around the body more efficiently – potentially providing a boost to the brain.

Moving your weight around also strengthens the muscles, adding to a sense of confidence and self-esteem and, when moving forwards, takes you physically and mentally to a better place.

2) Synchronisation

Humans are social creatures and our movements provide a powerful way to bond with others, particularly when done in groups. Brain-imaging studies have found that students working in groups begin to synchronise their brainwave patterns when they collaborate, and early research hints that the same happens when we dance. We already know that moving to music synchronises individual brains to the beat, and that moving together blurs the boundaries between 'self' and 'other', making people more likely to collaborate. So it makes sense to do some form of movement that involves synchrony, whether it be dance, drumming, tai chi or a group exercise class. All these will potentially do the same thing – make us feel connected to each other. Or, if you are physically alone and feeling adrift, getting into the 'groove' by moving to music (even just nodding along) can help to bring the world just a little bit closer.

3) Survival skills

You don't have to swim across rivers, climb trees for coconuts or start throwing spears at unsuspecting rabbits, but taking your body through the kind of movements it was built for makes you feel pretty good. The emerging links between fascia, stretching and mobility and a health immune system suggest that moving in the full human range of motion will keep fluids moving through the body as they should, in ways that could keep mood-sapping inflammation at bay.

This could involve slow, fluid motions, like a gentle swim or taking a good stretch-and-limber to prepare the joints to move. Or more explosive running, jumping and throwing – movements that release stored energy and frustrations all in one go. In times of stress, making use of our species' well-developed throwing skills is a great way to let off steam. If you don't have a dog to throw sticks for and don't fancy baseball or cricket, there are now places you can go to learn to throw axes for fun. Try it, it might just help.

4) Belly–nose control

Not a new yoga move, but the ability to move your diaphragm at a rate of six breaths per minute, while breathing exclusively through the nose.

Whatever rate you breathe at, whether you focus on the breath for alertness, slow it down a bit for relaxation or a lot to reach an altered state of consciousness, only nasal breathing allows your brainwaves to synchronise with the breath, offering a fast track to an alternative state of mind.

Plus, there is evidence that inhaling deeply can help both focus and memory while increasing the amount of oxygen flowing around the body. Mouth breathing, on the other hand, avoids all these benefits and contributes to bad breath and tooth decay.

5) Body-mindedness

The need to take the mind out of the head and put it back in the body is a theme that links pretty much all movement research to improved overall well-being. It works because focusing on the body forces you into the present and brings attention to bodily sensations that may signify the need for action.

Research is starting to reveal that moving while focusing on the body provides many of the same benefits as high-intensity exercise but, unlike Zumba, circuits or high-intensity interval training (HIIT), can be done by anyone, young, old, able-bodied or less so.

This suggests that some form of slow, quiet and deliberate movement that is done with the express intention of listening in to your body is an important foundation for any movement plan. It might sound a bit New Age to some, but it's essential to remind yourself that you are not a brain on legs, but a fully integrated mind–body beast. And that beast can't run on thoughts – or mindless iron-pumping – alone.

6) Free your mind

This is the exact opposite of embodiment, and involves unleashing the mind from the body and allowing yourself to just 'be'. Thanks to the strange-but-true world of brainwave entrainment, rhythmic movements are your friend. When our attention is fixed on the beat, the body moves along with very little conscious effort, and this temporarily unshackles the mental from the physical. The trance-inducing effect of moving to a beat was the way we got out of our minds before we had chemical alternatives, and it still works as well as it ever did.

For a lower-key version of the same thing, running or walking, or in fact any rhythmic repetitive movement – cycling, skiing or anything that you can do well enough to do without thinking – is as vital to our well-being as any other form of rest. It is the easiest way to use your movements to access the kind of creativity that usually shows up only at the most inconvenient times, like when we are in the shower or drifting off to sleep. Do it alone, let your mind wander and marvel at the weird and wonderful things that bubble up.

7) Learning through doing

As Peter Lovatt found when he applied dance skills to reading, physical literacy can extend way beyond the ability to move well: it can translate into new ways of thinking. Our culture tends to equate learning with sitting, but we are built to learn on the move. By taking our bodies through the movements they were made for, we open our minds to new ways of understanding the world and what we can achieve within it.

However you get your movement, feeling strong, agile and in control of your body is a potent source of self-confidence and belief, an antidote to anxiety and a short-cut to feeling better in general. Whether you get there via greater strength, balance or rhythmic movement, the implicit knowledge that your body is fit for life is well worth the time and effort.

*

Now for the difficult part: finding room in your life to fit it all in. Given that we're all rushed off our feet, perhaps the easiest way to do this is to forget finding more time to exercise and instead build more movement into your everyday life.

One way to do this is to take inspiration from the rare examples of people around the world who, unlike the rest of us, aren't slothing themselves into the physical, psychological and cognitive skids. There are five longevity hotspots around the world where ten times more people than average make it to a hundred years old. In all of them – Sardinia, Ikaria island in Greece, Okinawa in Japan, Nicoya in Costa Rica and Loma Linda in California – people are far less likely than average to suffer dementia and mental health issues. And, importantly, they are rarely found sitting around.

On the other hand, they rarely engage in what you might call 'exercise'. Instead, low-level activity such as gardening, foraging and walking just happens to be a part of their day. This, after all, is what we were built for. The Hadza, a group of modern hunter–gatherers in Tanzania who live, more or less, as our ancestors did, don't exercise either. The men average around 11 kilometres per day, hunting with bows and arrows, and climbing trees to collect honey. The women walk around 6 kilometres and dig into dry earth for tubers with a sharpened stick.[1] It's not easy work, but it's not a HIIT session either. Studies by the evolutionary anthropologist Herman Pontzer revealed that the Hadza use roughly the same number of calories per day as the average Westerner. They just use them smarter. When they do squats, it's not to feel the burn but to take a rest without getting a dusty backside. In fact, their legs don't even ache, because they are so used to standing up from hovering near the floor. And they live as long as we do, are healthier physically and, as one recent study noted wanly, 'seem happier than the Western scientists who visit them'.[2]

Low-level, all day movement, then, seems to be close to the ideal scenario for both body and mind – keeping the mental and physical cogs oiled, and blood, lymph and all our other bodily fluids moving on the inside in ways that support thinking, feeling and moving. But that doesn't change the fact that it's not easy to achieve when you sit to commute to a desk job and then sit all the way home again to flop on the sofa.

So, what to do? Standing desks and even desks with treadmills or exercise bikes attached are starting to make an appearance in more enlightened businesses. Walking meetings are also becoming a thing, but that only really helps if you're senior enough to push for them – and for the kind of meeting that doesn't require note-taking. As for choosing to stand while watching TV, or walking around during the adverts, if you can keep to that for more than one evening then you're a stronger person than me.

Changing the habits of a lifetime is hard. And, as the behaviour-change psychologist Theresa Marteau, of the Behaviour and Health Research Unit at the University of Cambridge, recently told me, the only environment we really have control over is the home.[3] Psychology studies suggest that we make so many of our decisions more or less in our sleep – reacting to unconscious cues and doing without thinking. Which means that the only way to bring more movement into everyday life is to change your home environment to the point where you can't help but move more.

One option is to go furniture-free, a trend that is having its moment in certain circles. Following in the (bare) footsteps of movement guru Katy Bowman, movement enthusiasts everywhere are replacing their sofas with cushions on the floor and sawing the legs off their dining tables so that they can squat or kneel instead of sitting. Or, for those who want the look without the student flat vibe, you can even spend hundreds of dollars on a designer desk that is the perfect height either to squat at or to pop on top of a normal desk to convert it into a standing one.

There's no need to go that far, unless you really want to, but there is no doubt that spending more time on the floor is probably a good thing to aim for. The reason is simple: at some point you are going to have to get up, which provides the equivalent of leg-pressing your entire body weight each time, which can't fail to increase leg strength. Constantly hopping up and down all day will also improve balance, slowing the loss of stability that comes with middle age.

This is clearly easier if you work at home and are free to move, fidget, sit on the floor or even squat in front of your computer without anyone giving you funny looks. I do all of the above, plus, being 4 feet 11 inches, my feet rarely reach the floor when I'm sitting anyway, and are usually tucked under me in one way or another, which lends itself to both fidgeting and regularly changing positions.

One rule of thumb, from a fairly recent study of sedentary behaviour and health, is that if you can't fidget like this, and movement isn't an integral part of your job, you should probably aim to get up and move every twenty to thirty minutes.[4] For the average person, who reads at the rate of 250 words per minute, that means stretching your legs or taking a dance break every ten pages or so. Remember, sitting all day is still bad for you, even if you are exercising regularly. The answer isn't to do more exercise overall, but to move little and often.

The MovNat crew have come up with a veritable buffet of 'movement snacks' which they suggest you sprinkle throughout the day. Like edible snacks, these tend to add up without you noticing and, before you know it, can make a measurable difference to your body. MovNat-approved snacks include crawling on all fours and in an upward-facing crab position, getting up from the floor (and back down again) without using your hands, balancing on one leg or hanging from the doorway by your fingertips. But anything can count as a movement snack as long as you get up and do something.

With enough fidgeting, walking around the block and perhaps a bit of gardening or housework, adding more movement into your life doesn't necessarily need to mean squeezing yet more activities into your day. It might even mean spending less time at the gym for a better all-round result. And it might also have the added benefit of cutting out that nagging feeling that you should probably be meditating to improve focus and mental health. By tuning in to your body more often you're being mindful and tuning out from the chatter of your thoughts by default. Job done. Even better, if you can make movement a fundamental feature of everyday life, you need never feel guilty about being still again.

And Finally: A Movement Manifesto

I truly believe that the findings outlined in this book have the power not only to boost our health as individuals but also to change our society for the better.

There is a groundswell of people who are learning and teaching that movement works, and a parallel body of science showing how and why it's worthwhile. But, let's face it, if the past few years have taught us anything, it's that facts and knowledge aren't necessarily enough to engender change, and as things stand we are a long way from a life founded on mind-boosting movement.

While there are many great projects on the go around the world, the fact remains that many of the people who could benefit the most aren't getting the resources necessary to make it happen. Bringing movement back into everyday life is going to need a serious investment of time, money and effort. Here, for what it's worth, is where I think change is most needed, and how the people with the means to make this happen can do the most good.

1) Start 'em young

The nineteenth-century American social reformer Frederick Douglass said, 'It's easier to build strong children than mend broken men'. Modern psychology agrees with him. So it follows that the best way to put movement back at the heart of our lives is first to accept that we have been screwing it up for some time now, and then put serious resources behind putting that right for the next generation.

Given that as many as 20 per cent of young people suffer from mental health issues, and the findings about the importance of interoception in developing emotional literacy, a priority should be to build some form of mindful movement into the school day. In very young children this could easily be wrapped up in imaginary play. In a recent trial by researchers at the Center for Healthy Minds at the University of Wisconsin-Madison, four- and five-year-olds were asked to imagine they were various animals by moving like an elephant or a snail. The group leader would gently draw their attention to bodily sensations as they did so, asking things like 'Can you feel your breath filling up the shell of the snail?' or 'Can you squeeze the elephant's trunk really hard to stop the water coming out and then let it go?' As far as the children were concerned, it was just a fun game, but it seems to have had a big impact. In a study published in the journal *Developmental Psychology* in 2015, the Wisconsin team reported that these interventions, along with other forms of mindfulness-based training, led to improvements in pro-social behaviour and emotional development compared with controls, as well as an improvement in their report-card grades.[1]

Seven- to eleven-year-olds and secondary-school-age children might be less enamoured with this approach. Even so, the benefits of mindful movement are even more important at this age, as children begin to navigate a more complex social world and become ever more self-sufficient. At this age the same body-reading skills could be delivered via martial arts or yoga-based lessons, or even more left-field options: what about circus skills, or parkour? And there's no reason why basic sporting skills, such as how to transfer weight to put more power behind a ball, couldn't also be taught through the medium of mindful movement.

Whether this is done by bringing in specialist teachers or as part of physical education lessons, investment is clearly needed to make this happen. So is a fresh look at the status given to physical education (PE) in schools. In the UK, PE is compulsory, but schools get to choose how much of the recommended minimum of two hours per week they can squeeze into the timetable.[2] Quite apart from the fact that two hours a week is almost nothing, up to a third of secondary schools have made cuts to PE time to make room for exam-cramming.

In the US there are no country-wide standards, and according to Andy Milne, an award-winning physical and health education teacher and blogger based in Illinois, physical activity is all too often seen as surplus to requirements. 'We're seeing more and more pressure for PE time to be cut or even removed,' he told me. 'Schools can scrap it either because they don't have the facilities or the teacher or because they think too much academic time is being curtailed.'

Given what we know about the importance of movement for mental health and cognition – not to mention physical health – this is worryingly short-sighted. Physical education researcher Jo Harris, of Loughborough University, made this exact point in 2019, when she called for PE to be considered a core subject, with the same weighting as English, maths and science. 'It is the only school subject whose primary focus is on the body,' she points out. Rather than being side-lined, says Harris, physical competence should be 'valued as much as reading, writing and arithmetic'.

As well as revitalising PE, schools could also take a leaf out of Elaine Wyllie's book. In 2012, when she was head teacher of a Scottish primary school, she realised that seden-tary lifestyles were affecting children's physical and mental well-being. She instigated the 'Daily Mile' – where children would down tools once a day and jog or run laps of the school playground for fifteen minutes. Class teachers could choose when to do it – perhaps when the children were looking particularly bored or restless – and the whole class would run together, each child deciding their own pace and making as much noise as they liked.

The Daily Mile has since spread to 11,000 schools and over 2 million children. A study from 2020 of over 5,000 children who regularly did the Daily Mile found that they achieved better scores on tests of cognition and well-being than a group who did a more intensive bleep test exercise, or those who stood around outdoors for fifteen minutes.[3]

For anyone wondering why children need to be told to run around when that is what playtime is for, the depress-ing answer is that playtime isn't the stalwart of the school day that it used to be. In the US, recess – break time – has long been a victim of cuts, with up to 40 per cent of school districts having cut recess time since the year 2000.[4] Figures from the US Centers for Disease Control estimate that schools average just twenty-seven minutes of recess time per day, and not all schools – even at elementary level – have scheduled break time at all.[5] In the UK it's a similar story. Some 85 per cent of UK primary schools and 50 per cent of infant schools have dropped afternoon playtime, and lunchtimes are becoming shorter than ever. According to the psychologist Ed Baines, who led the study that generated these figures, 'children barely have enough time to queue up and to eat their lunch, let alone have time for other things like socialising, physical exercise, or exploring self-chosen activities'.[6]

Meanwhile in Finland children have a fifteen-minute break after every forty-five-minute lesson, adding up to more than an hour of break-time per day, in which they are encouraged to be physically active. And they still achieve some of the best academic results in the world.

It doesn't matter how you look at it, moving more is good for kids – not just emotionally and physically but academically too. We need to turn this around, and the sooner the better.

2) Body-based therapy

It's one thing to pay lip-service to the benefits of physical activity for mental health, quite another to bring that properly into standard treatment regimes. In our brain-centric world, giving people the embodied knowledge that thoughts are not the only route to feelings is going to be hugely important in any serious attempt to improve our mental health. To be clear, I have nothing against medication, meditation or talk therapy: all three have helped me at one time or another. But movement needs to be given equal weighting as a way to take control of the mind – and prescribed in just the same way.

There is already some great stuff out there, for those who look for it. Some doctors in the UK prescribe 'Green Gyms' – conservation projects run by volunteers with the aim of combining environmental projects with improving both physical and mental health. At the last count there were over a hundred such schemes around the UK, and anecdotal reports point to improved well-being and relief from symptoms of mental ill health.

Given that talk therapies are expensive, with long waiting lists, and that in the US they may not be covered by health insurance, these kinds of schemes are hugely important. As we've seen, the evidence for strength-building, stretching, dance and breath control for mental health is unambiguous. It can be done in groups – indeed it works better in groups – and can hit many of the different pillars of mental well-being at once. It's time for the medical profession – and those who fund it – to stop treating movement as an optional extra and put some serious money behind movement-based treatments and therapies.

The news is somewhat better for people with learning difficulties – at least in special education settings, where the value of movement and physical experience is well understood, and core exercises, movement breaks and sensory awareness are part and parcel of the school day.

Since so many children with special needs are in mainstream education where movement is increasingly side-lined, however, there is room for improvement. And adults with special needs could also benefit from better funding of movement-based therapy. Charities and grassroots organisations are doing great work in this regard, but provision varies in different areas and funding is not easy to come by. My hope is that, as we start to see well-being for all through the lens of a mind–body perspective, practices that are specialist right now will become both common and easy to access.

3) Choose movement

You're an adult, and it's not my place to tell you what to do. But all the evidence suggests that adulthood, and especially middle age, is about the worst time to stop moving. The 'use it or lose it' nature of body and mind is never truer than in middle age: this is the time of life when the body gets busy gobbling up spare capacity in muscles, bone and brain, and what you don't use will only get harder to replace.

There is no one catch-all way that we should all move, no 'one weird trick' to make clickbait out of. What works for me might bore you to tears and vice versa. But if you're hitting the right movement 'buttons' outlined on pp. 193–7, then you won't be going far wrong. The only thing that we all need to do the same is to realise that making time for movement in your life is not a luxury, and it's not self-indulgent: it's a necessity. It'll lift mood, increase focus to get the boring stuff done faster, better and with fewer meltdowns while relieving stress and reminding you who you are. And, with a bit of luck, it might just put something in the bank for a healthier and happier old age.

4) Elderly action

The final group that is crying out for movement interventions is the elderly. There are often good physical – and social – reasons why older people are sedentary up to 80 per cent of the time. But there are also good reasons why movement, in whatever capacity is physically possible, makes a difference to physical, mental and emotional health. This is especially true in conditions like Parkinson's disease, where dance has been shown to relieve both physical and emotional symptoms, and dementia, where it can bring comfort, familiarity and a sense of community to people who might otherwise feel lost and alone.

Tai chi, walking, gardening and seated exercise classes are also great ways to improve strength, balance and confidence. This has been shown time and again, yet these services are still underfunded, and all too often left to charities to keep going. I can't put it any better than Terry Kvasnik, who has also led movement classes – even breakdancing – with groups of elderly people. 'There's no funding in it but [...] I feel like giving movement practices back to people is one of the least things we can do,' he says.

In the end, I hope that I have been able to persuade you that, whatever your age and whatever life has thrown at you so far, there is nothing that can't be made better by a short, sharp movement break.

Now, if you'll excuse me, there's a dance party about to start in my kitchen.

Notes

Introduction

1. Hoffmann, B., Kobel, S., Wartha, O., Kettner, S., Dreyhaupt, J., and Steinacker, J. M., 'High sedentary time in children is not only due to screen media use: a cross-sectional study', *BMC Pediatrics*, 2019, vol. 19(1): 154.
2. Harvey, J. A., Chastin, S. F., and Skelton, D. A., 'How sedentary are older people? A systematic review of the amount of sedentary behavior', *Journal of Aging and Physical Activity*, 2015, vol. 23(3): 471–87.
3. Bakrania, K., Edwardson, C. L., Khunti, K., Bandelow, S., Davies, M. J., and Yates. T., 'Associations between sedentary behaviours and cognitive function: cross-sectional and prospective findings from the UK biobank', *American Journal of Epidemiology*, 2018, vol. 187(3): 441–54.
4. Colzato, L. S., Szapora, A., Pannekoek, J. N., and Hommel, B., 'The impact of physical exercise on convergent and divergent thinking', *Frontiers in Human Neuroscience*, 2013, vol. 7: 824.
5. Smith, L., and Hamer, M., 'Sedentary behaviour and psychosocial health across the life course', in *Sedentary Behaviour Epidemiology*, ed. Leitzmann, M. F., Jochem, C., and Schmid, D., Springer Series on Epidemiology and Public Health (New York: Springer, 2017).
6. Teychenne, M., Costigan, S. A., and Parker K., 'The association between sedentary behaviour and risk of anxiety: a systematic review', *BMC Public Health*, 2015, vol. 15: 513. Zhai,

    L., Zhang, Y., and Zhang, D., 'Sedentary behaviour and the risk of depression: a meta-analysis', *British Journal of Sports Medicine*, 2015, vol. 49(11): 705–9.

7. Smith and Hamer, 'Sedentary behaviour and psychosocial health across the life course'.

8. Haapala, E. A., Väistöa, J., Lintua, N., Westgate, K., Ekelund, U., Poikkeus, A.-M., Brage, S., and Lakka, T. A., 'Physical activity and sedentary time in relation to academic achievement in children', *Journal of Science and Medicine in Sport*, 2017, vol. 20: 583–9.

9. Biddle, S. J. H., Pearson, N., Ross, G. M., and Braithwaite, R., 'Tracking of sedentary behaviours of young people: a systematic review', *Preventive Medicine*, 2010, vol. 51: 345–51.

10. Falck, R. S., Davis, J. C., and Liu-Ambrose, T., 'What is the association between sedentary behaviour and cognitive function? A systematic review', *British Journal of Sports Medicine*, 2017, vol. 51(10): 800–11.

11. Lynn, R., 'Who discovered the Flynn effect? A review of early studies of the secular increase of intelligence', *Intelligence*, 2013, vol. 41(6): 765–9.

12. Dutton, E., der Linden, D., and Lynn, R., 'The negative Flynn Effect: a systematic literature review', *Intelligence*, 2016, vol. 59: 163–9.

13. Lynn, R., 'New evidence for dysgenic fertility for intelligence in the United States', *Social Biology*, 1999, vol. 46: 146–53.

14. Rindermann, H., and Thompson, J., 'The cognitive competences of immigrant and native students across the world: an analysis of gaps, possible causes and impact', *Journal of Biosocial Science*, 2016, vol. 48(1): 66–93.

15. Ng, S. W., and Popkin, B. M., 'Time use and physical activity: a shift away from movement across the globe', *Obesity Reviews*, 2012, vol. 13: 659–80.

16. Claxton, G., *Intelligence in the Flesh: Why Your Mind Needs Your Body Much More Than It Thinks* (New Haven, CT: Yale University Press, 2015).

Notes

Chapter 1: Why We Move

1. Llinás, R. R., *I of the Vortex: From Neurons to Self* (Cambridge, MA: MIT Press, 2001).
2. Barton, R. A., and Venditti, C., 'Rapid evolution of the cerebellum in humans and other great apes', *Current Biology*, 2014, vol. 24: 2440–44.
3. Halsey, L. G., 'Do animals exercise to keep fit?', *Journal of Animal Ecology*, 2016, vol. 85(3): 614–20.
4. Lieberman, D. *The Story of the Human Body: Evolution, Health and Disease* (New York: Pantheon Books, 2013).
5. Raichlen, D. A., and Alexander, G. E., 'Adaptive capacity: an evolutionary neuroscience model linking exercise, cognition and brain health', *Trends in Neurosciences*, 2017, vol. 40 (7): 408–21.
6. Osvath, M., 'Spontaneous planning for future stone throwing by a male chimpanzee', *Current Biology*, 2007, vol. 19(5): 190–91.
7. Raby, C. R., Alexis, D. M., Dickinson, A., and Clayton, N. S., 'Planning for the future by western scrub-jays', *Nature*, 2007, vol. 445: 919–21.
8. Held, R., and Hein, A., 'Movement-produced stimulation in the development of visually guided behavior', *Journal of Comparative and Physiological Psychology*, 1967, vol. 56 (5): 872–6.
9. O'Regan, J. K., *Why Red Doesn't Sound like a Bell* (New York: Oxford University Press, 2011).
10. Humphrey, N., 'Why the feeling of consciousness evolved', *Your Conscious Mind: Unravelling the Greatest Mystery of the Human Brain*, New Scientist Instant Expert series (London: John Murray, 2017), pp. 37–43.
11. Craig, A. D., 'How do you feel – now? The anterior insula and human awareness', *Nature Reviews Neuroscience*, 2009, vol. 10(1): 59–70.

Chapter 2: The Joy of Steps

1. http://darwin-online.org.uk/EditorialIntroductions/vanWyhe_notebooks.html

2. Raichlen, D. A., and Alexander, G. E., 'Adaptive capacity: An evolutionary neuroscience model linking exercise, cognition and brain health', *Trends in Neurosciences*, 2017, vol. 40(7): 408–21.

3. Raichlen, D. A., Foster, A. D., Gerdeman, G. L., Seillier, A., and Giuffrida, A., 'Wired to run: exercise-induced endocannabinoid signaling in humans and cursorial mammals with implications for the "runner's high"', *Journal of Experimental Biology*, 2012, vol. 215: 1331–6.

4. Lee, D. Y., Na, D. L., Seo, S. W., Chin, J., Lim, S. J., Choi, D., Min, Y. K., and Yoon, B. K., 'Association between cognitive impairment and bone mineral density in postmenopausal women', *Menopause*, 2012, vol. 19(6): 636–41.

5. Berger, J. M., Singh, P., Khrimian, L., Morgan, D. A., Chowdhury, S., Arteaga-Solis, E., Horvath, T. L., Domingos, A. I., Marsland, A. L., Yadav, V. K., Rahmouni, K., Gao, X.-B., and Karsenty, G., 'Mediation of the acute stress response by the skeleton', *Cell Metabolism*, 2019, vol. 30: 1–13.

6. https://www.ambrosiaplasma.com

7. https://www.fda.gov/BiologicsBloodVaccines/SafetyAvailability/ucm631374.htm

8. https://onezero.medium.com/exclusive-ambrosia-the-young-blood-transfusion-startup-is-quietly-back-in-business-ee2b7494b417

9. Source: aabb.org (Blood FAQ: 'Who donates blood?' [accessed 16 August 2020]).

10. Lakoff, G., and Johnson, M., *Metaphors We Live By* (Chicago, IL: Chicago University Press, 1980).

11. Miles, L. K., Karpinska, K., Lumsden, J., and Macrae, C. N., 'The meandering mind: vection and mental time travel', *PLoS One*, 2010, vol. 5(5): e10825.

12. Aksentijevic, A., and Treider, J. M. G., 'It's all in the past: deconstructing the temporal Doppler effect', *Cognition*, 2016, vol. 155: 135–45.

13. Yun, L., Fagan, M., Subramaniapillai, M., Lee, Y., Park, C., Mansur, R. B., McIntyre, R. S., Faulkner, G. E. J., 'Are early

increases in physical activity a behavioral marker for successful antidepressant treatment?', *Journal of Affective Disorders*, 2020, vol. 260: 287–91.

14. Michalak, J., Troje, N. F., Fischer, J., Vollmar, P., Heidenreich, T., and Schulte, D., 'Embodiment of sadness and depression – gait patterns associated with dysphoric mood', *Psychosomatic Medicine*, 2009, vol. 71(5): 580–87.

15. Michalak, J., Rohde, K., Troje, N. F., 'How we walk affects what we remember: gait modifications through biofeedback change negative affective memory bias', *Journal of Behavior Therapy and Experimental Psychiatry*, 2015, vol. 46: 121–5.

16. Darwin, F., *Rustic Sounds, and Other Studies in Literature and Natural History* (London: John Murray, 1917).

17. Dijksterhuis, A., and Nordgren, L. F., 'A theory of unconscious thought', *Perspectives on Psychological Science*, 2006, vol. 1(2): 95–109.

18. Dijksterhuis, A., 'Think different: the merits of unconscious thought in preference development and decision making', *Journal of Personality and Social Psychology*, 2004, vol. 87(5): 586–98.

19. Chrysikou, E. G., Hamilton, R. H., Coslett, H. B., Datta, A., Bikson, M., and Thompson-Schill, S. L., 'Noninvasive transcranial direct current stimulation over the left prefrontal cortex facilitates cognitive flexibility in tool use', *Cognitive Neuroscience*, 2013, vol. 4(2): 81–9.

20. For a full account of this experiment see my previous book, *Override* (London: Scribe, 2017). Published in the US as *My Plastic Brain* (Buffalo, NY: Prometheus, 2018).

21. Oppezzo, M., and Schwartz, D. L., 'Give your ideas some legs: the positive effect of walking on creative thinking', *Journal of Experimental Psychology: Learning, Memory, and Cognition*, 2014, vol. 40(4): 1142–52.

22. Plambech, T., and Konijnendijk van den Bosch, C. C., 'The impact of nature on creativity – a study among Danish creative

professionals', *Urban Forestry & Urban Greening.* 2015, vol. 14 (2): 255–63.

23. https://www.ramblers.org.uk/advice/facts-and-stats-about-walking/participation-in-walking.aspx

24. Bloom, N., Jones, C. I., Van Reenen, J., and Webb, M., Are Ideas Getting Harder To Find? Working Paper 23782, National Bureau of Economic Research, 2017. https://www.nber.org/papers/w23782

Chapter 3: Fighting Fit

1. Barrett Holloway, J., Beuter, A., and Duda, J. L., 'Self-efficacy and training for strength in adolescent girls', *Journal of Applied Social Psychology*, 1988, vol. 18(8): 699–719.

2. Fain, E., and Weatherford, C., 'Comparative study of millennials' (age 20–34 years) grip and lateral pinch with the norms', *Journal of Hand Therapy*, 2016, vol. 29(4): 483–8.

3. Sandercock, G. R. H., and Cohen, D. D., 'Temporal trends in muscular fitness of English 10-year-olds 1998–2014: an allometric approach', *Journal of Science and Medicine in Sport*, 2019, vol. 22(2): 201–5.

4. https://www.ncbi.nlm.nih.gov/pmc/articles/PMC5068479/

5. Damasio, A., *The Feeling of What Happens: Body, Emotion and the Making of Consciousness* (London: Vintage, 2000), p. 150.

6. Barrett, L., *Beyond the Brain: How Body and Environment Shape Animal and Human Minds* (Princeton, NJ: Princeton University Press, 2011), p. 176.

7. Damasio, *The Feeling of What Happens.*

8. Alloway, R. G., and Packiam Alloway, T., 'The working memory benefits of proprioceptively demanding training: a pilot study', *Perceptual and Motor Skills,* 2015, vol. 120(3): 766–75.

9. Van Tulleken, C., Tipton, M., Massey, H., and Harper, C. M., 'Open water swimming as a treatment for major depressive disorder', *BMJ Case Reports* 2018, article 225007.

10. O'Connor, P. J., Herring, M. P., and Caravalho, A., 'Mental

health benefits of strength training in adults', *American Journal of Lifestyle Medicine*, 2010, vol. 4(5): 377–96.

11. Roach, N. T., and Lieberman, D. E., 'Upper body contributions to power generation during rapid, overhand throwing in humans', *Journal of Experimental Biology*, 2014, vol. 217: 2139–49.

12. https://youtu.be/HUPeJTs3JXw?t=2585 The crouch-somersault-crouch segment happens at 43.05 mins.

13. Schleip, R., and Müller, D. G., 'Training principles for fascial connective tissues: scientific foundation and suggested practical applications', *Journal of Bodywork and Movement Therapies*, 2013, vol. 17(1): 103–15.

14. Bond, M. M., Lloyd, R., Braun, R. A., and Eldridge, J. A., 'Measurement of strength gains using a fascial system exercise program', *International Journal of Exercise Science*, 2019, vol. 12(1): 825–38.

15. https://uk.news.yahoo.com/brutal-martial-art-saved-complex-114950334.html.

16. Van der Kolk, B. A., and Fisler, R., 'Dissociation and the fragmentary nature of traumatic memories: overview and exploratory study', *Journal of Traumatic Stress*, 1995, vol. 8(4): 505–25.

17. Janet, P., *Psychological Healing: A Historical and Clinical Study* (London: Allen and Unwin, 1925).

18. Rosenbaum, S., Sherrington, C., and Tiedemann, A., 'Exercise augmentation compared with usual care for post-traumatic stress disorder: a randomized controlled trial', *Acta psychiatrica scandinavica*, 2015, vol. 131(5): 350–59; Rosenbaum, S., Vancampfort, D., Steel, Z., Newby, J., Ward, P. B., and Stubbs, B., 'Physical activity in the treatment of post-traumatic stress disorder: a systematic review and meta-analysis', *Psychiatry Research*, 2015, vol. 230(2): 130–36.

19. Gene-Cos, N., Fisher, J., Ogden, P., and Cantrell, A., 'Sensorimotor psychotherapy group therapy in the treatment of

complex PTSD', *Annals of Psychiatry and Mental Health*, 2016, vol. 4(6): 1080.

20. Ratey, J., and Hagerman, E., *Spark! How Exercise Will Improve the Performance of Your Brain* (London: Quercus, 2008), p. 107.

21. Mukherjee, S., Clouston, S., Kotov, R., Bromet, E., and Luft, B., 'Handgrip strength of World Trade Center (WTC) responders: the role of re-experiencing posttraumatic stress disorder (PTSD) symptoms', *International Journal of Environmental Research and Public Health*, 2019, vol. 16(7): 1128.

22. Clouston, S. A. P., Guralnik, J., Kotov, R., Bromet, E., and Luft, B. J., 'Functional limitations among responders to the World Trade Center attacks 14 years after the disaster: implications of chronic posttraumatic stress disorder', *Journal of Traumatic Stress*, 2017, vol. 30(5): 443–52.

Chapter 4: Slave to the Rhythm

1. Phillips-Silver, J., Aktipis, C. A., and Bryant, G. A., 'The ecology of entrainment: foundations of coordinated rhythmic movement', *Music Perception*, 2010, vol. 28(1): 3–14.

2. Source: https://www.statista.com/statistics/756629/dance-step-and-other-choreographed-exercise-participants-us/#statisticContainer

3. Aviva UK Health Check Report, spring 2014.

4. Hanna, J. L., 'Dancing: a nonverbal language for imagining and learning', *Encyclopedia of the Sciences of Learning*, ed. Seel, N. M. (Boston, MA: Springer, 2012).

5. Neave, N., McCarty, K., Freynik, J., Caplan, N., Hönekopp, J., and Fink, B., 'Male dance moves that catch a woman's eye', *Biology Letters,* 2011, vol. 7(2), 221–4.

6. At the rock shelters of Bhimbetka in central India.

7. Winkler, I., Háden, G. P., Ladinig, O., Sziller, I., and Honing, H., 'Newborn infants detect the beat in music', *PNAS*, 2009, vol. 106(7): 2468–71.

8. Lewis, C., and Lovatt, P. J., 'Breaking away from set patterns of

thinking: improvisation and divergent thinking', *Thinking Skills and Creativity*, 2013, vol. 9: 46–58.

9. Gebauer, L., Kringelbach, M. L., and Vuust, P., 'Ever-changing cycles of musical pleasure: the role of dopamine and anticipation', *Psychomusicology: Music, Mind, and Brain*, 2012, vol. 22(2): 152–67.

10. Bengtsson, S. L., Ullén, F., Ehrsson, H. H., Hashimoto, T., Kito, T., Naito, E., Forssberg, H., and Sadato, N., 'Listening to rhythms activates motor and premotor cortices', *Cortex*, 2009, vol. 45(1): 62–71.

11. MacDougall, H., and Moore, S., 'Marching to the beat of the same drummer: the spontaneous tempo of human locomotion', *Journal of Applied Physiology*, 2005, vol. 99: 1164.

12. Moelants, D., 'Preferred tempo reconsidered', *Proceedings of the 7th International Conference on Music and Cognition*, ed. Stevens, C., Burnham, D., McPherson, G., Schubert, E., and Renwick, J. (Adelaide: Causal Productions, 2002).

13. Fitch, W. T., 'The biology and evolution of rhythm: unravelling a paradox', *Language and Music as Cognitive Systems*, ed. Rebuschat, P., Rohmeier, M., Hawkins, J. A., and Cross, I. (Oxford: Oxford University Press, 2011).

14. Patel, A. D., Iversen, J. R., Bregman, M. R., and Schulz, I., 'Experimental evidence for synchronization to a musical beat in a nonhuman animal', *Current Biology*, 2008, vol. 19(10): 827–30. Snowball, the dancing cockatoo: https://www.youtube.com/watch?v=N7IZmRnAo6s

15. Tarr, B., Launay, J., and Dunbar, R. I. M., 'Music and social bonding: 'self-other' merging and neurohormonal mechanisms', *Frontiers in Psychology*, 2014, vol. 5: 1096.

16. McNeill, W. H., *Keeping Together in Time: Dance and Drill in Human History* (Cambridge, MA: Harvard University Press, 1995).

17. Cirelli, L., Wan, S. J., and Trainor, L. J., 'Fourteen-month-old infants use interpersonal synchrony as a cue to direct

helpfulness', *Philosophical Transactions of the Royal Society, B*, 2014, vol. 369(1658).

18. Janata, P., Tomic, S. T., and Haberman, J. M., 'Sensorimotor coupling in music and the psychology of the groove', *Journal of Experimental Psychology*, 2012, vol. 141: 54.

19. Honkalampi, K., Koivumaa-Honkanen, H., Tanskanen, A., Hintikka, J., Lehtonen, J., and Viinamäki, H., 'Why do alexithymic features appear to be stable? A 12-month follow-up study of a general population', *Psychotherapy and Psychosomatics*, 2001, vol. 70: 247.

20. Di Tella, M., and Castelli, L., 'Alexithymia and fibromyalgia: clinical evidence', *Frontiers in Psychology*, 2013, vol. 4: 909.

21. Jeong, Y., and Hong, S., 'Dance movement therapy improves emotional responses and modulates neurohormones in adolescents with mild depression', *International Journal of Neuroscience*, 2005, vol. 115: 1711.

22. Bojner Horwitz, E., Lennartsson, A-K, Theorell, T. P. G., and Ullén, F., 'Engagement in dance is associated with emotional competence in interplay with others', *Frontiers in Psychology*, 2015, vol. 6, article 1096.

23. Campion, M., and Levita, L., 'Enhancing positive affect and divergent thinking abilities: play some music and dance', *Journal of Positive Psychology*, 2013, vol. 9: 137.

24. Spoor, F., Wood, B., and Zonneveld, F., 'Implications of early hominid labyrinthine morphology for evolution of human bipedal locomotion', *Nature*, 1994, vol. 23: 645.

25. Todd, N., and Cody, F., 'Vestibular responses to loud dance music: a physiological basis of the "rock and roll threshold"?', *Journal of the Acoustic Society of America*, 2000, vol. 107: 496.

26. Todd, N., and Lee, C., 'The sensory-motor theory of rhythm and beat induction 20 years on: a new synthesis and future perspectives', *Frontiers in Human Neuroscience*, 2015, vol. 9, article 444.

Notes

Chapter 5: Core Benefits

1. Pilates, J., and Miller, J. M., *Return to Life through Contrology* (New York: J. J. Augustin, 1945).
2. Middleton, F. A., and Strick, P. L., 'Anatomical evidence for cerebellar and basal ganglia involvement in higher cognitive function', *Science* 1994, vol. 266: 458–61.
3. Tallon-Baudry, C., Campana, F., Park, H. D., and Babo-Rebelo, M., 'The neural monitoring of visceral inputs, rather than attention, accounts for first-person perspective in conscious vision', *Cortex*, 2018, vol. 102: 139–49.
4. Stoffregen, T. A., Pagulayan, R. J., Bardy, B. B., and Hettinger, L. J., 'Modulating postural control to facilitate visual performance', *Human Movement Science*, 2000, vol. 19 (2): 203–20.
5. From WHO: https://www.who.int/news-room/fact-sheets/detail/falls
6. Balogun, J. A., Akindele, K. A., Nihinlola, J. O., and Marzouk, D. K., 'Age-related changes in balance performance', *Disability and Rehabilitation*, 1994, vol. 16(2): 58–62.
7. Wayne, P. M., Hausdorff, J. M., Lough, M., Gow, B. J., Lipsitz Novak, L. V., Macklin, E. A., Peng, C.-K., and Manor, B., 'Tai chi training may reduce dual task gait variability, a potential mediator of fall risk, in healthy older adults: cross-sectional and randomized trial studies', *Frontiers in Human Neuroscience*, 2015, vol. 9: 332.
8. Feldman, R., Schreiber, S., Pick, C. G., and Been, E., 'Gait, balance and posture in major mental illnesses: depression, anxiety and schizophrenia', *Austin Medical Sciences*, 2020, vol. 5(1): 1039.
9. Carney, D. R., Cuddy, A. J., and Yap, A. J., 'Power posing: brief nonverbal displays affect neuroendocrine levels and risk tolerance', *Psychological Science*, 2010, vol. 21(10): 1363–8.
10. https://faculty.haas.berkeley.edu/dana_carney/pdf_My%20position%20on%20power%20poses.pdf
11. Jones, K. J., Cesario, J., Alger, M., Bailey, A. H., Bombari, D.,

Move!

Carney, D., Dovidio, J. F., Duffy, S., Harder, J. A., van Huistee, D., Jackson, B., Johnson, D. J., Keller, V. N., Klaschinski, L., LaBelle, O., LaFrance, M., Latu, I. M., Morssinkhoff, M., Nault, K., Pardal, V., Pulfrey, C., Rohleder, N., Ronay, N., Richman, L. S., Schmid Mast, M., Schnabel, K., Schröder-Abé, M., and Tybur, J. M. Power poses – where do we stand?', *Comprehensive Results in Social Psychology*, 2017, vol. 2(1): 139–41.

12. Osypiuk, K., Thompson, E., and Wayne, P. M., 'Can tai chi and qigong postures shape our mood? Toward an embodied cognition framework for mind–body research', *Frontiers in Human Neuroscience*, 2018, vol. 12, article 174; https://www.ncbi.nlm.nih.gov/pmc/articles/PMC5938610/pdf/fnhum-12–00174.pdf.

13. Kraft, T. L., and Pressman, S. D., 'Grin and bear it: the influence of manipulated facial expression on the stress response', *Psychological Science*, 2012, vol. 23(11): 1372–8.

14. Wagner, H., Rehmes, U., Kohle, D., and Puta, C., 'Laughing: a demanding exercise for trunk muscles', *Journal of Motor Behaviour*, 2014, vol. 46(1): 33–7.

15. Weinberg, M. K., Hammond, T. G., and Cummins, R. A., 'The impact of laughter yoga on subjective well-being: a pilot study', *European Journal of Humour Research*, 2014, vol. 1 (4): 25–34.

16. Bressington, D., Mui, J., Yu, C., Leung, S. F., Cheung, K., Wu, C. S. T., Bollard, M., and Chien, W. T., 'Feasibility of a group-based laughter yoga intervention as an adjunctive treatment for residual symptoms of depression, anxiety and stress in people with depression', *Journal of Affective Disorders*, 2019, vol. 248: 42–51.

17. Schumann, D., Anheyer, D., Lauche, R., Dobos, G., Langhorst, J., and Cramer, H., 'Effect of yoga in the therapy of irritable bowel syndrome: a systematic review', *Clinical Gastroenterology and Hepatology*, 2016 vol. 14(12): 1720–31.

18. Liposcki, D. B., da Silva Nagata, I. F., Silvano, G. A., Zanella, K., and Schneider, R. H., 'Influence of a Pilates exercise

program on the quality of life of sedentary elderly people: a randomized clinical trial', *Journal of Bodywork and Movement Therapies*, 2019, vol. 23(2): 390–93.

Chapter 6: Stretch

1. Langevin, H. M., and Yandrow, J. A., 'Relationship of acupuncture points and meridians to connective tissue planes', *The Anatomical Record*, 2002, vol. 269: 257–65.
2. Eyckmans, J., Boudou, T., Yu, X., and Chen, C. S., 'A hitchhiker's guide to mechanobiology', *Developmental Cell*, 2011, vol. 21(1): 35–47.
3. Langevin, H. M., Bouffard, N. A., Badger, G. J., Churchill, D. L., and Howe, A. K., 'Subcutaneous tissue fibroblast cytoskeletal remodeling induced by acupuncture: evidence for a mechanotransduction-based mechanism', *Journal of Cellular Physiology*, 2006, vol. 207(3): 767–74.
4. Di Virgilio, F., and Veurich, M., 'Purinergic signaling in the immune system', *Autonomic Neuroscience*, 2015, vol. 191: 117–23. See also: Dou, L., Chen, Y. F., Cowan, P. J., and Chen, X. P., 'Extracellular ATP signaling and clinical relevance', *Clinical Immunology*, 2018, vol. 188: 67–73.
5. Liu, Y. Z., Wang, Y. X., and Jiang, C. L., 'Inflammation: the common pathway of stress-related diseases', *Frontiers in Human Neuroscience*, 2017, vol. 11: 316.
6. Falconer, C. L., Cooper, A. R., Walhin, J. P., Thompson, D., Page, A. S., Peters, T. J., Montgomery, A. A., Sharp, D. J., Dayan, C. M., and Andrews, R. C., 'Sedentary time and markers of inflammation in people with newly diagnosed type 2 diabetes', *Nutrition, Metabolism and Cardiovascular Diseases*, 2014, vol. 24(9): 956–62.
7. Franceschi, C., Garagnani, P., Parini, P., Giuliani, C., and Santoro, A., 'Inflammaging: a new immune-metabolic viewpoint for age-related diseases', *Nature Reviews Endocrinology*, 2018, vol. 14(10): 576–90.
8. Kiecolt-Glaser, J. K., Christian, L., Preston, H., Houts, C.

R., Malarkey, W. B., Emery, C. F., and Glaser, R., 'Stress, inflammation, and yoga practice', *Psychosomatic Medicine*, 2010, vol. 72(2): 113–21.

9. Berrueta, L., Muskaj, I., Olenich, S., Butler, T., Badger, G. J., Colas, R. A., Spite, M., Serhan C. N., and Langevin, H. M., 'Stretching impacts inflammation resolution in connective tissue', *Journal of Cell Physiology*, 2016, vol. 231(7): 1621–7.

10. Serhan, C. N., and Levy, B. D., 'Resolvins in inflammation: emergence of the pro-resolving superfamily of mediators', *Journal of Clinical Investigation*, 2018, vol. 128(7): 2657–69.

11. Benias, P. C., Wells, R. G., Sackey-Aboagye, B., Klavan, H., Reidy, J., Buonocore, D., Miranda, M., Kornacki, S., Wayne, M., Carr-Locke, D. L., and Theise, N. D., 'Structure and distribution of an unrecognized interstitium in human tissues', *Scientific Reports,* 2018, vol. 8(1): 4947.

12. https://www.researchgate.net/blog/post/interstitium

13. Panchik, D., Masco, S., Zinnikas, P., Hillriegel, B., Lauder, T., Suttmann, E., Chinchilli, V., McBeth, M., and Hermann, W., 'Effect of exercise on breast cancer-related lymphedema: what the lymphatic surgeon needs to know', *Journal of Reconstructive Microsurgery*, 2019, vol. 35(1): 37–45.

14. If you want to know if you're hypermobile, do the test here: https://www.ehlers-danlos.com/assessing-joint-hypermobility/

15. Eccles, J. A., Beacher, F. D., Gray, M. A., Jones, C. L., Minati, L., Harrison, N. A., and Critchley, H. D., 'Brain structure and joint hypermobility: relevance to the expression of psychiatric symptoms', *British Journal of Psychiatry*, 2012, vol. 200(6): 508–9.

16. Mallorquí-Bagué, N., Garfinkel, S. N., Engels, M., Eccles, J. A., Pailhez, G., Bulbena, A., Critchley, H. D., 'Neuroimaging and psychophysiological investigation of the link between anxiety, enhanced affective reactivity and interoception in people with joint hypermobility', *Frontiers in Psychology,* 2014, vol. 5: 1162.

17. https://www.medrxiv.org/content/10.1101/19006320v1

18. Mahler, K. *Interoception, the Eighth Sensory System* (Shawnee, KS: AAPC Publishing, 2016).

Chapter 7: Breathless

1. Iyengar, B. K. S., *Astadala Yogamala*, vol. 2 (New Delhi: Allied Publishers, 2000), p. 37.
2. There are a few examples of basic breath control in human-raised apes, including Koko the gorilla, who learned to play the harmonica and recorder, plus a captive orang-utan named Bonnie who worked out how to whistle by copying her keepers. Neither showed any ambitions for world domination, however. See: Perlman, M., Patterson, F. G., and Cohn, R. H., 'The human-fostered gorilla Koko shows breath control in play with wind instruments', *Biolinguistics*, 2012, vol. 6(3–4): 433–44.
3. Li, P., Janczewski, W. A., Yackle, K., Kam, K., Pagliardini, S., Krasnow, M. A., and Eldman, J. L., 'The peptidergic control circuit for sighing', *Nature*, 2016, vol. 530(7590): 293–7.
4. Vlemincx, E., Van Diest, I., Lehrer, P. M., Aubert, A. E., and Van den Bergh, O., 'Respiratory variability preceding and following sighs: a resetter hypothesis', *Biological Psychology*, 2010, vol. 84(1): 82–7.
5. MacLarnon, A. M., and Hewitt, G. P., 'The evolution of human speech: the role of enhanced breathing control', *American Journal of Physical Anthropology*, 1999, vol. 109(3): 341–63.
6. Heck, D. H., McAfee, S. S., Liu, Y., Babajani-Feremi, A., Rezaie, R., Freeman, W. J., Wheless, J. W., Papanicolaou, A. C., Ruszinkó, M., Sokolov, Y., and Kozma, R., 'Breathing as a fundamental rhythm of brain function', *Frontiers in Neural Circuits*, 2017, vol. 10: 115. Tort, A. B. L., Brankačk, J., and Draguhn, A. Respiration-entrained brain rhythms are global but often overlooked. *Trends in Neurosciences*, 2018, vol. 41(4): 186–97.
7. Arshamian, A., Iravani, B., Majid, A., and Lundström, J. N., 'Respiration modulates olfactory memory consolidation in

humans', *The Journal of Neuroscience*. 2018, vol. 38(48): 10286–94.

8. Zaccaro, A., Piarulli, A., Laurino, M., Garbella, E., Menicucci, D., Neri, B., and Gemignani, A., 'How breath-control can change your life: a systematic review on psycho-physiological correlates of slow breathing', *Frontiers in Human Neuroscience*, 2018, vol. 7(12): 353.

9. Bernardi, L., Sleight, P., Bandinelli, G., Cencetti, S., Fattorini, L., Wdowczyc-Szulc, J., and Lagi, A., 'Effect of rosary prayer and yoga mantras on autonomic cardiovascular rhythms: comparative study', *BMJ*, 2001, vol. 323(7327): 1446–9.

10. Bernardi, L., Spadacini, G., Bellwon, J., Hajric, R., Roskamm, H., and Frey, A. W., 'Effect of breathing rate on oxygen saturation and exercise performance in chronic heart failure', *Lancet*, 1998, vol. 351(9112): 1308–11.

11. Chung, S. C., Kwon, J. H., Lee, H. W., Tack, G. R., Lee, B., Yi, J. H., and Lee, S. Y., 'Effects of high concentration oxygen administration on n-back task performance and physiological signals', *Physiological Measurement*, 2007, vol. 28(4): 389–96.

12. Noble, D. J., and Hochman, S., 'Hypothesis: pulmonary afferent activity patterns during slow, deep breathing contribute to the neural induction of physiological relaxation', *Frontiers in Physiology*, 2019, vol. 13(10): 1176.

13. Yasuma, F., and Hayano, J., 'Respiratory sinus arrhythmia: why does the heartbeat synchronize with respiratory rhythm?', *Chest*, 2004, vol. 125(2): 683–90.

14. Payne, P., and Crane-Godreau, M. A., 'Meditative movement for depression and anxiety', *Frontiers in Psychiatry*, 2013, vol. 4, article 71.

Chapter 8: And … Stop
1. Often misattributed to Banksy.
2. Khan, Z., and Bollu, P. C., 'Fatal familial insomnia', *StatPearls* (Treasure Island, FL: StatPearls Publishing, 2020).
3. Fultz, N. E., Bonmassar, G., Setsompop, K., Stickgold, R.

A., Rosen, B. R., Polimeni, J. R., and Lewis, L. D., 'Coupled electrophysiological, hemodynamic, and cerebrospinal fluid oscillations in human sleep', *Science,* 2019, vol. 366(6465): 628–31.

4. Besedovsky, L., Lange, T., and Born, J., 'Sleep and immune function', *Pflugers Arch.,* 2012, vol. 463(1): 121–37.

5. Recommended amount of sleep for a healthy adult: a joint consensus statement of the American Academy of Sleep Medicine and Sleep Research Society, *Sleep,* 2015, vol. 38(6): 843–4.

6. Hammond, C., and Lewis, G., 'The rest test: preliminary findings from a large-scale international survey on rest', *The Restless Compendium: Interdisciplinary Investigations of Rest and Its Opposites,* ed. Callard, F., Staines, K., and Wilkes, J. (London: Palgrave Macmillan, 2016).

7. Hammond, C., *The Art of Rest: How to Find Respite in the Modern Age* (Edinburgh: Canongate, 2019).

Summary: Move, Think, Feel

1. Pontzer, H., Raichlen, D. A., Wood, B. M., Mabulla, A. Z. P., Racette, B., and Marlowe, F. W., 'Hunter–gatherer energetics and human obesity', *PLoS One,* 2012, vol. 7(7): e40503.

2. Reid, G., 'Disentangling what we know about microbes and mental health', *Frontiers in Endocrinology,* 2019, vol. 10: 81.

3. Williams, C., 'How to trick your mind to break bad habits and reach your goals', *New Scientist,* 24 July 2019.

4. Diaz, K. M., Howard, V. J., Hutto, B., Colabianchi, N., Vena, J. E., Safford, M. M., Blair, S. N., and Hooker, S. P., 'Patterns of sedentary behavior and mortality in U.S. middle-aged and older adults: a national cohort study', *Annals of Internal Medicine,* 2017, vol. 167(7): 465–75.

And Finally: A Movement Manifesto

1. Flook, L., Goldberg, S. B., Pinger, L., and Davidson, R. J., 'Promoting prosocial behavior and self-regulatory skills in

preschool children through a mindfulness-based kindness curriculum', *Developmental Psychology*, 2015, vol. 51(1): 44–51.

2. https://www.education-ni.gov.uk/articles/statutory-curriculum#toc-2

3. Booth, J. N., Chesham, R. A., Brooks, N. E., Gorely, T., and Moran, C. N., 'A citizen science study of short physical activity breaks at school: improvements in cognition and well-being with self-paced activity', *BMC Medicine*, 2020, vol. 18(1): 62.

4. https://www.cdc.gov/healthyschools/physicalactivity/pdf/Recess_Data_Brief_CDC_Logo_FINAL_191106.pdf

5. https://www.cdc.gov/healthyschools/physicalactivity/pdf/Recess_All_Students.pdf

6. https://www.ucl.ac.uk/ioe/news/2019/may/break-time-cuts-could-be-harming-childrens-development

Acknowledgements

The idea for this book took root on one of many long, ponderous dog walks. The fact that it grew to be something more is thanks to the many people who listened to my ramblings, shared their knowledge and experience, and encouraged me to dig deeper. I am especially grateful to my agent, Peter Tallack, for some crucial pointers early on and for his dogged determination to make the book a reality.

I am also indebted to my editors, Ed Lake at Profile and John Glynn at Hanover Square Press, who have been hugely enthusiastic and supportive throughout and offered much-needed votes of confidence throughout the writing and editing process. Thanks, too, to Matthew Taylor for being gentle while combing out the tangles.

Without the scientists who so generously gave up their time to talk to me about their work and what it means, I wouldn't have much to write about. Huge thanks must go to Peter Strick, Peter Wayne, Eric Kandel, Gerard Karsenty and Rebecca Barnstaple, who allowed me to invite myself to their labs and offices and were incredibly generous with their time and expertise. Also, thanks to Helene Langevin, Dennis Muñoz-Vergara, Dick Greene, Peter Lovatt, David Raichlen, Hugo Critchley, Jessica Eccles, Neil Todd, Petr Janata, Micah Allen, Richard Dunn, David Levinthal and Elizabeth Broadbent for some fascinating and revealing conversations along the way.

Equally important were the inspirational people who not only shared their stories and experiences with me but are doing incredible work in helping people discover how movement can change their lives. Marcus Scotney, Terry Kvasnik, Jerome Rattoni, Sharath Jois,

Move!

Hamish Hendry, Andy Milne, Dale Youth boxing club and Kevin Edward Turner and his wonderful dance group – thank you all for your help – and keep up the good work!

Finally, love and thanks to my friends and family for putting up with me and for asking 'How's the book going?' when they knew the answer would be more than they bargained for. Thanks to Anna, Will and George for putting me up (again!) in Boston, and to Iain and Jess for looking after us in New York. And finally, to Jon, Sam and Jango, my favourite three fidgets, who never sit still if they can possibly help it. Thank you for forcing me up and out, even when I can't be bothered. You, quite literally, keep me moving, and I couldn't do any of it without you.

Index

Index

Index

groove enhancement machine (GEM) 93
growth factors 33
gut 114, 122, 130, 145, 146, 175
gyrotonics 131

H
Hadza people 19–20, 198–9
Hammond, Claudia: *The Art of Rest* 185
'happy' movements 105–6
Harris, Jo 205–6
Harvard Business School 116–17
Harvard Medical School 144
Harvard University 133, 135, 137
Hava Nagila 105–6
heart rate variability (HRV) 177, 178
Hébert, Georges 63–4
Herrero, Jose 168, 169
high-intensity exercise 6, 188, 196
high intensity interval training (HIIT) 196, 198
hippocampus 19, 33, 38, 57
homeostasis 25, 58, 74
Homo erectus 20
hunter–gatherers 19–20, 30–2, 198
hunting skills 18, 19, 30–2, 68, 198
hypermobility 149–53, 154, 156
hypofrontality 48, 49–50

I
irritable bowel syndrome (IBS) 130, 150
imagination 23
immune system 134, 135, 139–41, 143, 145, 146, 147, 154, 156, 176, 184, 188, 195
inflammation 8, 40, 139, 140–5, 176, 178, 184, 187, 188, 195
insula (area of cortex) 25
interoception 24–6, 27, 59, 70, 150, 153, 168, 188, 204
IQs, sedentary lifestyles linked to falling 4, 5–6, 30
Iyengar, B. K. S. 157

J
Janata, Petr 93, 94–6
Janet, Pierre: *Psychological Healing* 74
Johnson, Mark 44
Jois, K. Pattabhi 148, 155
Jois, Sharath 148, 149, 155, 164, 168
jumping 63, 64, 67, 68–71, 79, 80, 105, 109, 195

K
Kandel, Eric 36, 38, 39
Karsenty, Gerard 37, 38–40, 41
Kolk, Bessel van der 73–4, 75–6
Krav Maga 72
Kringelbach, Morten 88
Kung Fu 24, 69

Index

motor cortex 121, 125, 126
motor neurons 67
movement
  evolution of 11–27
  low-level, all day 198–9
  manifesto 203–10
  plan, must-have elements of
    193–7
  snacks 201
  see also individual area of
    movement
MovNat 62–6, 201
muscle
  core 112, 114, 115, 116, 127,
    128, 139, 130–1, 132
  strength 17–18, 57, 58, 59, 63,
    65, 66–8, 71, 79, 112, 114,
    115, 116, 127, 128, 139,
    130–1, 132
  stretching 133–5, 147, 148,
    152–3, 154, 155–6
muscular bonding 92
musculoskeletal division 58, 59
music 2, 62, 68, 81, 83–4, 88,
  89, 90, 91, 92, 93, 94–5, 96–7,
  103, 106, 108, 180, 186, 194
Music Perception 81

N

nasal breathing 165–7, 171, 182,
  195
National Bureau of Economic
  Research 51
natural exercise 62–6, 79
Nāṭya Śāstra (Hindu text) 99

Nazi Party 94
Neanderthals 160
nervous system 11–12, 24, 40,
  71, 75, 112, 123, 124, 128,
  151, 155, 172, 176, 181
Nietzsche, Friedrich 29, 30
9/11 attacks 77

O

obesity 142
Ogden, Pat 73, 76
O'Regan, J. Kevin 23
osteoblasts 37
osteocalcin 36–41, 54, 193
osteoporosis 36
otoliths 106, 108
oxygen saturation, blood 174–5

P

pandiculation 133–4, 154
panic attack 72, 170
parkour 55, 63, 205
Payne, Peter 179, 180
Pilates 9, 111–12, 113, 114, 130,
  131, 132
Pilates, Joseph 111
planning ability 4, 16–17, 89,
  164
Pontzer, Herman 198
post-traumatic stress disorder
  (PTSD) 72–4, 75–6
posture 45, 59, 101, 112–13,
  114, 115–21, 128, 132, 148,
  168
power posing 117, 118

Index

Index